Country Faith

A New England Country Village Church

Country Faith

*Rural America Stories of Faith from Forty of the
Most Inspirational People of Our Time*

Volume I

*AMERICA'S CHRISTIAN HERITAGE &
MESSAGES OF FAITH SERIES*

BRENDA ERVIN

Rural America Books

Cover Photo of the Billy Graham Library compliments of Billy Graham Evangelistic Association, 1 Billy Graham Parkway, Charlotte, North Carolina 28201. Call 704-401-3200 for tour information. www.billygrahamlibrary.com

Cover Photo: Cave Springs Presbyterian Church, Missouri
Photo Credit: Denny Medley, Denny Medley/Random Photography, Springfield, Missouri
www.randomphotography.com

Beaumont Ranch and the Chisholm Fork Cowboy Chapel, © 2009 Linda Beaumont
Liberty University Worley Chapel © 2009 Liberty University
Little Brown Church in the Vale, © 2009 Vicki Mann
The Old Rugged Cross Historic Site by Brenda Ervin
Old Rugged Cross Church © 2009 Brenda Ervin
All other photos property of Rural America Books

Library of Congress Control Number: 2007921042
ISBN: 978-0-9674808-0-0

Ervin, Brenda, Country Faith ©2009 First Edition

Project Director: Celeste Bennett McCormick
Cover Designer: Barry Kerrigan
Interior Book Design: Sue Balcer of JustYourType.biz

Unless noted all scripture references come from the Amplified Bible

Includes Bibliographical References: Spiritual/Agricultural Geography:
    1. Religious 2. Religious Leaders 3. Agricultural   5. Regional/United States Geography
    6. Inspirational 7. Vernacular Architecture

Published by Rural America Books
In association with Barn Door Publishing
Web: www.ruralamericabooks.org
Email: countryfaith@comcast.net

All rights reserved. No part of this book may be reproduced or transmitted in any form or by any means without written permission from author under penalty of United States Copyright Law. This book is to educate, entertain and honor the subject matter. Every effort humanly possible has been extended to make this book accurate, however, there may be mistakes both typographical and in content. Any unacknowledged material in this book may be sent to the publisher and proper credit will be given in future editions. The author does not represent the Billy Graham Evangelistic Association, nor is she a spokesperson for any of the ministries or organizations contained in this book.

*Dedicated to these inspiring leaders
early in my spiritual life:*

Billy Graham

Lowell Lundstrom

Oral Roberts

Pat Robertson

Kenneth & Gloria Copeland

Marilyn Hickey

Charles F. Stanley

"Thus says the Lord:

Stand by the roads and look; and ask for the eternal paths,

Where the good, old way is; then walk in it, and you will find

rest for your souls…"

Jeremiah 6:16

Contents

INTRODUCTION:
Safeguarding Our Agri-Spiritual Heritage 9

PART ONE:
Faith in the Country: Pioneer Preachers, Prayers & Protracted Revivals 18
Billye Brim 25
Charles Capps 26
George Washington Carver 27
Mark Chironna 28
Gloria Copeland 29
Kenneth Copeland 31
Paul Crouch Sr. 33
The Little Brown Church in the Vale 35
Jerry Falwell, Sr. 41
Jentzen Franklin 42
A Farm Boy's Song: Alfred E. Brumley 43
Bill Gaither 44
The Mother Church of Country Music:
The Ryman & Reverend Samuel Porter Jones 45

PART TWO:
God's Work & The American Barn 50
Ruth Graham 56
Franklin Graham 57
The Old Rugged Cross Hymn Historic Site: Featuring George Beverly Shea 60
Blackie Gonzales 66
John Hagee 67
Kenneth Hagin, Sr. 68
Kenneth Hagin, Jr. 69
Marilyn Hickey 71
Benny Hinn 73
Rex Humbard 74

PART THREE:
Our Sacred Sanctuaries: Featuring Dexter Avenue Memorial Baptist Church 78
T. D. Jakes 86
Coretta Scott King 87
Martin Luther King, Jr 88
Rosa Parks 89

PART FOUR:
Faith of a Quaker Miller: Greatest Biblical Wheat Experiment in the World 94
Henry Krause 100
F. Dewey Lockman 101
Lowell Lundstrom 103
Peter Marshall, Chaplain of U.S. Senate 104
Joyce Meyers 105
Keith Moore 106
J. E. Murdock 107
Mike Murdock 108
John Osteen 110
Dotti Rambo 111
Christian Farmer's Outreach: Facing the Future with Faithfulness 112

PART FIVE:
Faith on the Farm: Lessons of Sowing & Reaping, Featuring Oral Roberts 116
Pat Robertson 120
Cheryl Salem 122
Farm Rescue: Good Samaritan's of the Heartland 123
Jerry Savelle 124
R. W. Schambach 125
Robert Schuler 127
Demas Shakarian 129
Charles F. Stanley 130
Billy Sunday 131
Village Missions: Keeping Country Churches Open 132

PART SIX:
America's Cowboy Churches: Barns, Chapels & Cowboy Faith 140
The Dineh, the Navajo: Desert Camp Meetings, Miracles & Ministries 148

Acknowledgements 154
Endnotes 156
Bibliography 161

Introduction

Safeguarding Our Agri-Spiritual Heritage

…Should God ask you to do so…

~George Washington Carver

On a long, red dirt road in rural Mecklenburg County, North Carolina, a wounded veteran of The War Between the States, returned to his ancestral home and built a log cabin on his Sharon Township farmland. Born in 1840, William Crook Graham served with the 6th South Carolina Volunteers, 1st Company, and carried a Yankee bullet in his leg for the rest of his life. Known as a "hard drinking, hard cursing veteran," William Crook Graham was of Scots-Irish heritage. His devoutly Christian wife, Maggie McCall Graham, reared their eleven children to love God, and inherited from her the "teachings of scripture." Sons, William Franklin and Clyde Graham, inherited their father's devotion to the land. After their father passed away in 1910, they carried on their family's generational farm that at the time was one of 5,000 farms remaining in Mecklenburg County. William Franklin Graham and his wife Morrow Coffey Graham, and, Clyde Graham, managed the *Graham Farm Dairy's* three hundred acres.

On November 7, 1918, only eight years after the death of patriarch William Crook Graham, his grandson was born in the family clapboard farmhouse. William Franklin Graham and Morrow Coffey Graham named their newborn son, William Franklin Graham, Jr., who was later to become the greatest Gospel evangelist the world has ever known. His extraordinary life of faithfully serving God for more than 60-years is unequaled. Throughout his life he has preached to "more people in live audiences than anyone else in history, an estimated 215 million in more than 185 countries." Reared on his family's Mecklenburg County, North Carolina dairy farm, young

Billy Frank, as he was called, worked hard in the farm fields and inside the barns. Rising before dawn, he spent much of his youth walking the familiar pathway toward the white dairy barn to milk their seventy-five cows. Later on, he assisted in delivering milk on a dairy route into nearby Charlotte. Morrow Coffey Graham kept the books for the burgeoning dairy farm, while William Franklin and Clyde managed the day-to-day operations. According to records the Graham Farm Dairy was one of the earliest dairies in Sharon Township, a southern place where honeysuckle vines "looped and tangled over the farm lane fences."

But life on the farm was sometimes difficult. For example, Reverend Graham recalls in the 1920's, during a long dry spell, he watched his father's discouraged-looking face as he returned from waterless fields for "no rain meant no crops."

"I remember when milk went down to five cents a quart, how worried they were as to whether they could make it or not," said Reverend Graham in *Billy Graham: God's Ambassador*. People of prayer and strong faith in God, the Graham's trusted the Lord to provide for their farming operation no matter what the economic conditions. By 1927, when he was nine-years old, the Graham Farm Dairy on Park Road was thriving and a two-story Colonial-style brick farmhouse was built.

Today, the Graham Farmhouse is beautifully situated and safeguarded on sixty-five acres of the Billy Graham Evangelistic Association a few miles from the original farm location. Nearby, *The Billy Graham Library* stands near the Graham Farmhouse and created to reflect the image of the dairy barn of Reverend Graham's childhood. "It's an instrument, a tool for the Gospel," said Reverend Graham.

The Billy Graham Library was built by ITEC of Orlando, Florida, and began construction in late October 2005, substantially completed in December 2006. An undertaking of more than four years to design and create, the Billy Graham Library consists of a Main Barn and two Side Wings.

An awe-inspiring forty-foot cross captures visitors' attention on the front of the Main Barn, and is a visual reminder of the redemptive Cross of Jesus Christ. "The cross is made from six, eight- foot square, one-half inch thick tinted glass panels and a matching pair of eight foot tall by four foot wide entry doors," says Jeff Burton, President of ITEC of Orlando, Florida, and designer of the state-of-the-art library.

Opened to the public on June 5, 2008, the Main Barn's interior is of durable southern yellow pine with a stained finish, and is approximately 48 feet wide by 73 feet deep by 50 feet, honoring the vast woodlands environs of Reverend Graham's youth. Campbellsville Industries, Kentucky, specializing in steeples, was chosen to fabricate the aluminum cupolas for the Main Barn, each adding an additional 7 feet, 6 inches in height. Beside the Main Barn, the two Barn Wings are about 34 feet wide and 63 feet deep, and approximately 21 feet high, says Jeff Burton. Inside the Billy Graham Library's Main Barn bookstore, the 20-inch wide ponderosa pine columns and beams adds to the distinctive quality of this one-of-a-kind 40,000 square foot structure. The Main Barn's exterior siding is stained western red cedar, which compliments the landscape, and nearby brick Graham Farmhouse, referred to as the *Graham Family Homeplace*.

A concrete Silo, approximately 61 feet 6 inches to the top, is covered with natural North Carolina stacked stone veneer wall base, says designer, Jeff Burton.

Visitors sense the years of devotion and love of the Graham family with memorabilia from the Billy Graham Crusades and, through 350 photographs and films, discover the historic journey of "America's Ambassador." The Billy Graham Library also houses many farm related items of the Graham Farm Dairy, such as an old truck used on the farm. "Ruth's Attic," commemorates the life of author and poet, Ruth Bell Graham, wife of Reverend Billy Graham.

Franklin Graham, son of Billy and Ruth Graham, and President of the Billy Graham Evangelistic Association, says, "It is my prayer that this center will be a testimony to the world of the amazing love of God and the great things He can do through lives that are surrendered to Him." Through the foresight and efforts of the Billy Graham Evangelistic Association (BGEA), this library "that faith built" offers a glimpse into the lives of this beloved North Carolina farm family whose Christian influence has transformed generations. This writer's love and respect for the Billy Graham Ministry dates back decades ago when growing up on a former dairy farm in rural southeast Michigan in the early 1960's.

"A heritage has been left to us by this generation…
I believe God is going to hold us responsible for what we do with our inheritance"
~J. Roswell Flower, Editor

Country Faith began more than a decade ago after discovering that many of our nation's most inspiring leaders grew up in rural America. As an agricultural writer specializing in farm family and barn heritage, this often-overlooked yet remarkable aspect of our nation's farm history fascinated me. After discovering many farm and barn stories of spiritual greats, such as Reverend Graham, I noticed a commonality between their stories; not only did they serve God in some aspect of ministry, but they all grew up close to the earth. Coming from all denominations, cultures and geographical backgrounds, I was captivated by their rural American stories—some were actually second and third generational farmers. For example, author Dr. J. E. Murdock, now in his nineties, farmed early in life; Demos Shakarian, a third generation California farmer owned the largest privately-owned dairy in the world; Kenneth Copeland worked alongside his grandfather on his Texas farm; John Osteen worked in cotton fields as a young man; Bill Gaither, a third generation farmers son worked on his family's Indiana dairy farm; and third-generation Arkansas farmer Charles Capps farmed until the 1970's.

Others you'll discover in this book worked the land, tended cattle, picked cotton, mended fences, milked cows, managed gardens and orchards. Such as, evangelist Marilyn Hickey who worked on a Pennsylvania apple farm, and Cheryl Salem, former Miss America, was reared on a Mississippi farm. Some drew close to God in backwoods solitude, like Pastor Keith Moore, or in farm fields and in the quietness of old barns, like Reverend Billy Graham. As a story-gatherer, I placed these culturally rich tales inside my favorite wooden farm-produce box. Later on, I submitted portions of my book to peers and friends. One of the most frequent comments was: "I didn't know so many grew up on farms!"

What would our nation be like without the influence of these icons of faith or their farm family's spirit?

I spent countless hours over the course of several years with this question on my mind. Like the treasured stories of those who've risen above the norm, such as the "Mother of the Civil Rights Movement," Rosa Parks, who picked cotton as a six-year old girl, Dr. Martin Luther King who was a sharecropper's son, and T.D. Jakes "America's Bishop," who managed a garden at a young age, their diverse and distinct experiences demanded my attention. Pain and isolation, loss and injustice, and *justice* as in the story of Dr. Jerry Savelle, their *agonia*, tests and trials endured on the way to greatness, changing the course of history were birthed in rural places. Often facing unimaginable hardships, many risked everything to reach their God-given goals.

For example, today most people have heard of Dr. Oral Roberts, founder of Oral Roberts University, Dr. Paul Crouch, founder of TBN, and Dr. Jerry Falwell, founder of Liberty University, but little is known or recognized of their pioneer ancestors who were farmers and endured many trials farming the land. Dr. Roberts worked on his family's Oklahoma farm like his preacher father before him, Dr. Crouch preserves his farm ancestors memory in his books, and Dr. Farwell's ancestors were pioneers and innovative Lynchburg, Virginia farmers.

As my idea for this book soon became an all-encompassing mission, I began to ask myself questions, like, *Why haven't these amazing stories been written about before?* My fascination soon evolved into a consuming passion. As a preservationist, journalist and a devotional writer, I believed as the Psalmist said, that God's faithfulness and handiwork within the lives of each of these people featured should be honored and remembered. Overtime, my curiosity was peaked enough to contact some sources and their responses disappointed me. I suppose it shouldn't shock us that America's spiritual and agricultural heritage is nowhere to be found on the list of the most popular topics by the media in our culture today.

As an agricultural author who talks to these dedicated and self-sacrificing Americans, that fact troubles me. True, there's no glitz or glamour in the barnyard. But what type of nation do we wish to be if all we care about are those sensational topics, often faddish, that do not sustain the soul or compel the human spirit to reach for eternity or eternal qualities?

While concentrating on my aim, this book, I soon gained support and encouragement from many. Professional courtesies came into my life beyond my wildest imagination, and doors opened for me that I could never have opened myself. I sensed God was overwhelmingly involved in this project. It was as if He was *asking me to do so,* as George Washington Carver once said.

"I think she has religion!"
~Michigan Farmer

Once a farmer friend took me to visit a neighboring farmer to document his barn and farm. After my farmer friend jumped out of the vehicle as soon as we arrived, he ran up to the barn where the man was waiting and I quickly followed. This is what I heard:

"Now don't be swearing around her Joe! I don't know what she has, but I think its *religion*!"

My journey of traveling more than two decades and 60,000 miles for my books, chronicling farm family heritage, has a history and a story of its own. Like this memory of one farm visit out of hundreds, my belief in my mission to gather farm family stories before they were gone was a personal journey and birthed by prayer. Before farm documentation was in the news I was committed to my vision that from the start was divinely orchestrated, and completed through persistence and determination. However, it actually began in the early 1960's as a young girl with a dream in my heart in the hayloft of an old dairy barn; *Who built the barns, why did they build them, what did they use them for, and why did the farm family leave the farm?* These were just a few of my musings as a pre-teen girl working in and around our barn, and that over time became larger than life to me. Near Rankin, Michigan, inside the barn, I'd write stories about my excursions in the countryside near century-old farms. While trekking through fallow pastures, I'd often think of the stirring messages from the Bible brought by a man on television called Billy Graham. Seeds for the new birth were placed into my young life. Years later other ministers profoundly effected me, such as a message by Minnesota-based Lowell Lundstrom, and messages from rural western churches also impacted my spiritual understanding. Through the ministries of those that I dedicate my book to, I found God's mercy and forgiveness.

Although barn and farm family preservation are two subjects I'm most recognized for both photographically and in my writings, all aspects of farm life remain the heart of my work. Quiet conversations in farm kitchens, on front porches, and inside barns, I listened to concerns over the loss of farmland as well as their faith in God. As a Christian devotional author, I felt deeply moved and connected with these stories of faith. Of course, not all farmers discussed their faith, but God was often brought up and a family Bible brought out for me to photograph. In *Barns of Michigan*, I wrote extensively about some of these stories such as the Bavarian missionary immigrants of Frankenmuth, Michigan. *The Old Rugged Cross Historic Church,* a former barn, and *The Good Shepherd Barn Mural* were also chronicled over the years and are featured in this book. Others such as, Grandpa requesting prayer at church during a drought, farmers with teams of horses and wagons gathering to build a church, or Alfred E. Brumley's story of his song, "I'll Fly Away," each story documented seemed to be leading and preparing me for this project.

"A heritage has been left to us by this generation, which is now passing away," wrote J. Roswell Flower in the 1940's. As true now as it was in J. Roswell Flower's time, our heritage is slowly eroding. Today its getting harder to experience rural America as projected numbers and growing trends toward development say it's never going to be the same again. America Farmland Trust says that 1.5 million of farm acres are lost every year. In fact, in Michigan if urban sprawl trends continue, more than 10 million acres will be gone by the year 2020. North Carolina lost 600,000 acres in just a few years, and Montana and other western states are witnessing similar losses. In addition, according to experts, farms, ranches, churches and sacred landmarks, are disappearing from many regions of the United States and my book preserves many of them for future generations.

Similar to many Americans, my ancestors were pioneer farmers. Faith in the country was just as important then as a whitewashed framed church, or an eye-beam in a timber-frame barn. Like, one Pottawatomie farm family ancestor who sung in a back roads country church in the Ojibwa language, for the Ojibwa people. Back then rural America was easy to find. Because of my heritage, small town values of faith, family and farm life have always been a large part of who I am and what I love. I care deeply about farm family and faith stories so much so that I've dedicated much of my life to preserve them, and write about them. They're not just another story to me, but my calling. What is important to them is important to me. I suppose this is what makes me different from other farm-related researchers and authors.

"Stand by the roads and look; and ask for the eternal paths, Where the good, old way is; then walk in it, and you will find rest for your souls…"
Jeremiah 6:16

"Safeguarding America's agri-spiritual heritage," is the theme of this book. Spiritual treasures like Old McKendree Chapel, near Cape Girardeau, Missouri, one of the first church histories obtained during the research and development of this project, along with many other agri-spiritual landmarks are introduced. My husband's ancestors, Hezekiah and Polly Ervin, homesteaded in the vicinity of Cape Girardeau County, near Jackson, Missouri. Originally from North Carolina, Hezekiah and Polly reared their family as Christians in southern Tennessee. Like many patriarchs of the day, Hezekiah Ervin wrote his children's births, deaths, and marriages in the family Bible. A cherished heirloom, the family Bible was passed down from generation to generation and protected. History records that oftentimes the family Bible was the only book in a pioneer's log cabin, and served as part of farm children's education. Within the pages of the Family Bible, instructions for life and for eternal life were read aloud to the family. Hezekiah wrote the following generational blessing in his Bible: *May all men born be born again In Christ to Live in Christ to reign, virtue and glory, and rejoice be their true unfading crown.*

As in the days of Hezekiah Ervin, and Old McKendree Chapel, today there are scores of Christian ministries still standing by the road and looking into the future and blessing generations to come. They are faithfully serving God and mankind, like, Brian Weschler, President of Village Missions of North America, Wilson Lippy, President of Christian Farmers Outreach, Susie Luchsinger and Russ Weaver hosts of Cowboy Church, RFD-TV, Jim Ballard's East Idaho Cowboy Church, Jeff Smith of Cowboy Church Network of North America and Bill Gross of Farm Rescue. Although the purpose of this book is to celebrate the agri-spiritual heritage of our spiritual greats and to bring awareness to these and other rural ministries, still, I believe it is vital to record some impacting our culture still, like, George Washington Carver, agricultural scientist, The Lockman Foundation, publisher of the Amplified Bible and other publications, and Henry Krouse, inventor of farm machinery are just a few of the agri-spiritual legacies I write about who have changed our world.

"Our nation is spiritually hungry," a CNN reporter once said. Over the years from a number of book lectures, I've discovered this statement to be very true. I found there is a deep hunger in the hearts of people for the way rural America used to be before disappearance of farmland, old churches and old barns—and also for the way it used to be *spiritually*. Similar to the biblical famine in Joseph's Egypt, I call this void a famine of the soul. A multi-faceted mission, one of the purposes of my book is to serve those who are famished for country faith, the way their parents or grandparents reared them, the old church where they first learned the scriptures, sang in the choir, sensed God speaking to their hearts for the first time, and grew to adulthood within the security of the sanctuary and community. In these pages I pray you will find peace, comfort, and you will behold the good old road, and, as the prophet Jeremiah said, find rest for your soul.

Cave Springs Memorial Church
Missouri

Part One

...they walked by faith and not by sight...

Hebrews 11: 1

Faith in the Country:

Pioneer Preachers, Prayers & Protracted Revivals

"Come, let us make it a subject of prayer under this haystack, while the dark clouds are going and the clear sky is coming."

~Samuel J. Mills

In August 1806 a group of college students in Williamstown, Massachusetts, gathered weekly for prayer near a grove of maples called Sloan's Meadow. Among the five young men was Samuel J. Mills, who kept a diary of this earnest, praying band of Christian brothers. One day while praying it began to "thunder and lighten," recalled one of the prayers, Byram Green. Leading the small group was Samuel J. Mills, who then cried out: "Come, let us make it a subject of prayer under this haystack." With faith and vision to reach Asia with the Gospel, this humble place of prayer became known as *The Haystack Prayer Meeting a*nd the birthplace of the American Foreign Missions that ultimately reached foreign lands.

"There came a revival of religion in New England," said Edward Warren Capen, historian of the American Foreign Missions Board. The Haystack Prayer Meeting touched hundreds of student's far beyond this farmer's field, and many committed their lives to missionary work. Adoniram Judson, one of the first Baptist foreign missionaries from America, was among the young men deeply changed by the Haystack Prayer Meeting. After consecrating his life to Christian ministry, with his missionary wife Anne, he was dedicated to "convert Burmans and the Burman Bible," wrote their son, Edward Judson. Sweeping through the land was a spirit of revival. Due to many factors, people were hungry for God in the early 19th century.

Years before, seeds of the salvation message were sown into the hearts of colonial America through the preaching of the Word by the some of the greatest evangelists of the day. According to historian Mark Noll, in 1734, Jonathan Edwards' sermon, "Justification by Faith Alone," stirred hearts within the Connecticut River Valley region that continued for years. "News circulated rapidly about the dozens, and then hundreds, who were 'brought to a lively sense of the excellency of Jesus Christ and his sufficiency and willingness to save sinners and to much weaned in their affections from the world,'" he says quoting Jonathan Edwards in *The Rise of Evangelicalism.*

America witnessed its first Great Awakening in the 1740's and English-born George Whitefield was instrumental in leading thousands to God through his preaching, including farmers. Alan Taylor, author and historian of *American Colonies: The Settling of North America*, records a Connecticut farmer's written conversion account: "I saw Mr. Whitefield…and my hearing him preach, gave me a heart wound; by God's blessing: my old Foundation was broken up."

Faith in the country was rising and the majority of people were converted in the colonies, says author Peter Marshall in *Sounding Forth the Trumpet*. Circuit-riding preachers were launched into the wilds of backwoods territories during this time, converting souls to Christianity. Although few in number, these ordained ministers were assigned by their denominational leaders, and committed to their calling, even if it required traveling great distances. Men of boldness, courage and perseverance, they ministered to the needs of pioneer farmers and their families, as well as to native bands of Indians known as First Peoples, bands of Indians in distant places. Prayers of the faithful did not return void as circuit riders and denominational pastors over the years answered the call to head west and evangelize. Mounted on their saddles, with only a Bible and few meager belongings, circuit riders endured tremendous hardships. Coming close to God, God came close to them.

> *"…Since the day of Pentecost there was hardly ever a greater revival of religion than at Cane Ridge."*
>
> ~Peter Cartwright, Circuit Rider Preacher

Circuit riders and denominational pioneer preachers were seasoned veterans of the uninhabited wilderness. They endured hardship as soldiers of Jesus Christ, and were experienced at carrying the Gospel to inhospitable places and pioneer settlements, like one carved from the Cumberland Mountains of Kentucky called Cane Ridge. Nothing could have prepared them for the massive spiritual awakening that was to take place within this Cumberland outpost. In August 1801, five years before The Haystack Prayer Meeting, one of the most awe-inspiring revivals in all history took place at Cane Ridge called *The Cane Ridge Revival.*

Before this time in 1796, Presbyterian Barton Warren Stone arrived and pastored the *Cane Ridge Presbyterian Church*, built in 1791. Regularly scheduled meetings in Kentucky, southern Ohio, and northern Tennessee brought Presbyterians together for fellowship, "They traveled to each other's sacramental communion services," says the Cane Ridge history. On the first Sunday of August 1801, Cane Ridge Presbyterian Church was the host of the gathering. Pastor Stone, and others, spread the word that it was going to be a "great meeting." Soon, thousands both "rich and poor" made the trek to the meeting, and within a few weeks the number grew to 12,000, recorded at the time by a military officer. Farmers dropped their plows and farmwomen left their chores. Loading their families in wagons, they rushed to the camp meeting, where all races and "every denomination" was represented.

Peter Cartwright, one of the most influential circuit-riding preachers of his time, wrote in his autobiography that thousands came by foot to Cane Ridge, as well as on horseback and in carriages and wagons. "From this camp meeting…news spread through all the Churches, and through all the land and it excited great wonder and surprise," he said. Circuit riders of the Methodist, Presbyterian and Baptist faiths flocked to preach and pray with earnest seekers encamped in tents, wagons, and under trees. Faith in God was alive.

Today, the *Cane Ridge Meeting House,* formerly the Cane Ridge Presbyterian Church, is recognized as the largest log structure in existence in the United States. Located near present-day Lexington, Kentucky, the place where the Cane Ridge Revival took place is lovingly preserved in a limestone enclosure, and local congregations, some with spiritual roots dating back to its revival days, including *The Churches of Christ,* presently are caretakers. After the Cane Ridge Revival, nothing would be the same in America.

Historian Jack Larkin, author of *The Reshaping of Everyday Life,* records the writings of one pioneer woman arrived at a camp meeting site, "In a wild district of Indiana…a space of about twenty acres had been cleared…in a circle round the cleared space.'" Camp meetings, the great open-air gatherings of the revival took worship outdoors and extended it to thousands at once, Jack Larkin says. Circles of tents and praying groups continued all across the territories and kept the revival spirit stirring until it eventually reached the southern states, says Anne C. Loveland in *Southern Evangelicals and the Social Order.*

"Come to Jesus"

~ American Tract Society

One of the greatest, yet least known protracted revivals in America, took place in the south during the Civil War. Author and Chaplain J. William Jones, known as the "Fighting Parson," was a respected spiritual leader of his day and recorded this unprecedented revival in his book, *Christ in the Camp.* In camps and on battlefields, chaplains, pastors of churches, and colporteurs (tract distributors ministering to spiritual needs) served the southern army. Antietam, Gettysburg, Fredericksburg, and Manassas, were just a few of the battlefields where conflicts took place on farms fields. Barns were also used for field hospitals where men of God ministered to the spiritually needy. Chaplain Jones witnessed dying soldiers holding Bibles and Gospel tracts in their hands, especially a tract called *Come to Jesus* as he ministered to General Lee's army. With more than 17,000 copies printed and selling for three-cents apiece, they were distributed to the "men in grey" and reached them around campfires and in tents, on battlefields and field hospitals. Men of an

Alabama regiment were converted by the tract "Come to Jesus," and a Mississippi regiment solider wrote that "Come to Jesus" had been the means of leading him to Christ since being in Virginia.

Founded in 1828, The American Tract Society distributed other tracts (small booklets) like *The Act of Faith* and *Don't Put it Off*, much asked for ministry tools. One soldier came upon a chaplain preaching and called out for one of his tracts, saying it was going to be used to keep him from any "devilment."

"Revivals of religion are contagious…The same gracious Heavenly Father that has owned and revived His work at Fredericksburg, and in other portions of the army, has at last poured out upon us refreshing showers of His grace," wrote Chaplain James Nelson, from the headquarters of the Forty-Fourth Virginia Regiment. Spiritually hungry, both those with Christian backgrounds returning to the Lord and those who never gave their life to God, all rested in the eternal promises of heaven for the Child of God in the Bible.

Chaplain Jones and his colleagues estimated that the number of conversions from this great awakening between 1861-1865 was upward to 100,000 men. "Was not Christ in the camp a vital power, and was not the camp a school for Christ?" testified Chaplain Jones. Prayers and preaching leading up to the revival and saturating camps kept the spirit of repentance, forgiveness of sins, through "coming to Jesus" for many years. Today restoration is underway to preserve the landmark *St. George Episcopal Church* in Fredericksburg, Virginia. Built in 1849, the church standing today is the third since the 1700s. By a close-up observation of this historic brick sanctuary not far from the Fredericksburg Battlefield, one is drawn to remember the consecrated lives of General Lee's troops it once sheltered.

"…The old pulpit consecrated by holy hands"

~Old McKendree Chapel

Seedtime and harvests continued and pioneers seeking a better life for themselves and their families traveled westward, to Missouri and other parts of the frontier territories. Near Jackson, Missouri, *Old McKendree Chapel* still stands after 190 years, and recognized as the oldest log structure west of the Mississippi. Built in 1818-1819, Cape Girardeau County's Old McKendree Chapel was a favored location for early regional conferences and prayer meetings conducted by some of the most esteemed southeast Missouri preachers of the time. Originally, the land on which the chapel was built was part of a farm, owned by pioneer and devout Methodist, William Willams. In 1806, the first camp meeting was held on this farm. With a nearby spring and "thicket of trees providing good shade," the Williams farm was a "good place" to congregate. "My great-great-great-great-grandfather, William Williams, donated the land upon which the chapel was built," says descendant Betty Poe Henry, Board of Trustees, whose ancestors have "kept the heritage close to their hearts."

By 1809, the McKendree Church was organized, and a year later Bishop William McKendree, fourth Methodist bishop and first born in the United States, took part in the camp meetings. In 1818 construction of the chapel began, and history records the chapel was "a good, hewed-log house," with whitewashed logs for pews. Built of hand-hewn poplar logs from the surrounding woods, the old pulpit was "consecrated by holy hands of bishops and blessed by the touch of humble preachers." In the 1920's restoration began to save the chapel that stands today on the original two-acres and renovated as a United Methodist Historic Shrine, and listed in the National Register of Historic Places in the United States. Over the years, hymns of faith such as, "Amazing Grace" and "Bringing In The Sheaves," were sung to give glory to God at Old McKendree Chapel and other country church's across the land.

Not far from Old McKendree in Greene County, Missouri, *Cave Springs Memorial Church* was the location of a nationally broadcast hymn singing directed and produced by country music artist Stan Hitchcock. Built in 1869, Cave Springs Memorial Church broadcast was inspired by a Bible that Stan Hitchcock's mother left him, "An old country church means to me memories of my mama, for she left me the gift of music." Founder of BlueHighways TV, Stan Hitchcock recorded Gospel classics like "Softly and Tenderly Jesus Is Calling," inside the acoustically-sound Cave Springs Memorial Church sanctuary along with Gospel artists friends Ronnie Reno, Barbara Fairchild and The Sullivan Family. Originally called Mt. Zion Presbyterian Church, it was renamed Cave Springs Memorial Church. Standing behind the rustic pulpit, the prayers of pioneer preachers of years past for souls to be saved and lives to be used of God from their church way out in the county were answered in ways they never dreamed possible. "The Country Church Songbook Collection," filmed at the Cave Springs Memorial Church is touching lives and bringing revival into the hearts and lives of people today.

"Changing the landscape of Christianity."

~A Los Angeles, California Barn & Revival

Pioneer means to go before. The American pioneer was dauntless as they journeyed on foot and horse and in wagons moving westward. Converted pioneers sought out other believers and established churches, businesses and farms. Pioneer preaching, powerful prayers of faith and protracted revivals went on across the land including those led by Billy Sunday, Mordecai Ham, and others of the early 20th century. Faith in God grew stronger in rural America. The fervent spirit and legacy of a pioneer preacher named William J. Seymour is still honored and revered. The son of an ex-slave, William J. Seymour, preached during a southern California revival that was held in a barn. This revival was later said by many to have "changed the landscape of Christianity." Born in 1870 in Centerville, Louisiana, he endured the painful humiliations of segregation. Once he was not permitted in the same room with other seminary students. However, little is much when God is in it, and his small acts of faithfulness were rewarded when he was later invited by a gathering of believers to conduct services in Los Angeles, California.

Beginning in April 1906, William J. Seymour delivered messages inside a two-story barn located at 312 Azuza Street, and people of all denominations gathered at the stable once called "a tumbled down shack." Within a short time, the small congregation became the place of one of the most significant protracted revivals the world has ever known. Much like former revivals in America, the *Azuza Street Revival* leaders and followers were persecuted for their beliefs. Even so, William J. Seymour's message spread to the African-American, Latino, and other communities, drawing them to this epicenter that eventually became known as the home of the Pentecostal Holiness Church and Pentecostal movement. The Azuza Street Revival (1906-1909) was often ignored, and the humble horse barn dismissed as a lost footnote to history. But in 2006 a celebration was held in honor of the Azuza Street Revival, with more than 45,000 from all over the globe attending the 100th anniversary celebration in Los Angeles, California.

Like William J. Seymour, pioneer preachers pressed on in faith disregarding innumerable obstacles and led prayer meetings and protracted revivals on battlefield farm fields, mountain settlements, log chapels, city horse stables and on dusty western cattle trails. Spiritual leaders throughout our nation's history, both known and unknown, founded ministries that continue to this day, including the American Foreign Missions Board, now called the International Missions Board, and the American Tract Society still strong after more than 180 years. Unto the Lord, they looked into the future with eyes of faith. Their stalwart endeavors to lift up the Gospel in rural places for the "good old way" (Jeremiah. 6:16) eventually helped to build the spiritual foundations of America.

While the earth remains, seedtime and harvest,
cold and heat, summer
and winter and day and night shall not cease.

~ Genesis 8:22

Billye Brim

"My great-great-grandmother, Lou (Jackman) Pipes, a widow, with her children farmed the land…Ike and Carrie Pickard (my grandparents) became known as praying people, and the Pickard Place was known as a place of prayer…Praying Pentecostals took refuge at the Pickard Place where a long deep canyon cut through their farm. And if they "had a burden" or "got happy" no critical ears could hear…I believe the spiritual canopy of blessing over the greater Tulsa area was because of these early saints and their faithfulness to pray… I have thoughtfully considered that Dr. Bill and Vonette Bright's Ministry, without their being aware of it, was possibly affected by such prayers…The Bright's lived on the farm next to my Grandpa and Grandma Coday and the families enjoyed much fellowship…William (Dr. Bill Bright, Campus Crusades for Christ Ministry) was a much younger brother."

~Billye Brim, Founder, Prayer Mountain of the Ozarks
Billye Brim Ministries

Author of The Road to Prayer Mountain

Charles Capps

"I could be the best rice farmer in the state of Arkansas, and I could sit in my house and say, 'Praise God, I believe in rice. My grandfather believed in rice. My daddy believed in rice'…I could have 10-tons of seed on my truck waiting to be planted, but if I just sit there and praise God because I believe in rice, I'll never harvest any rice…It's the Word and Faith that works."

~Charles Capps, Founder, Charles Capps Ministries, England, Arkansas
Third-Generation Arkansas Farmer

Author of, The Tongue: A Creative Force

George Washington Carver

"Keep your hand in that of the Master, walk daily by his side…then we can walk and talk with Jesus because we will be attuned to His will and wishes, thus making the Creation story of the world non-debatable as to its reality…I ask the Great Creator silently daily, and often many times per day, to permit me to speak to Him through the three great Kingdoms of the world, which he has created, the animal, mineral and vegetable Kingdoms; their relation to teach other, to us, our relation to them and the Great God who made all of us."

~George Washington Carver, Scientist

Excerpted from: George Washington Carver in his Own Words by Gary R. Kremer

Mark Chironna

"Possess the faith of God, make it your own, take the faith of God for yourself…No one can take it for you. God doesn't have any grandchildren; He only has sons and daughters. Lay hold of the faith of God; rise up in faith…Lift up your head and lay hold of the faith of God. You are a son of Almighty God."

~ Dr. Mark J. Chironna, Founder, Mark Chironna Ministries
Master's Touch International Church
Longwood, Florida

Author of *The Creative Authority of the Spoken Word*

Gloria Copeland

"Nearly thirty-five years ago Ken and I began living by faith. Ken was praying down in the riverbed in Tulsa, Oklahoma, and God began speaking to him about preaching to the nations…God said way back then we'd have a worldwide ministry. It was clear God hadn't considered our bank account. We hardly had enough money to get across town-much less go to the nations! One day Ken came home really excited and said, 'We are going to be Oral Roberts' Partners!' 'Where in the world will we get ten dollars to send to Oral Roberts?' I asked…As we began to sow seed we started to see increase… Since then God has given us stewardship over a ministry that takes millions of dollars to operate! And we still have to use our faith for all God calls us to do…God wants us to live by faith… I wouldn't want it any other way."

~Gloria Copeland, Co-Founder of Kenneth Copeland Ministries,
Fort Worth, Texas

Author of *God's Master Plan for Your Life*
Living by Faith and *The Unbeatable Spirit of Faith*

"Give ear and hear my voice; listen and hear my words.
Does he who plows for sowing plow continually?
Does he continue to plow and harrow the ground after it is smooth?
When he has leveled its surface, does he not cast abroad the seed of
dill or fennel and scatter cumin
(a seasoning) and put the wheat in rows, and barely in its intended place, and spelt (an
inferior kind of wheat) as the border?
…his God instructs him correctly and teaches him.
For dill is not threshed with a sharp threshing instrument,
nor is a cartwheel rolled over cumin; but dill is beaten off with a staff, and cumin with a
rod (by hand)…This also comes from the Lord of hosts, who is wonderful in counsel and
excellent in wisdom and effectual working."

~Isaiah 28: 23-29

Kenneth Copeland

"My Grandfather was a master farmer. Many times on his 640-acres, he had crops when the rest of the county didn't have any. One time I remember the only reason the cotton gin opened up was to gin my Grandfather's cotton because nobody else in the county had any. Sometimes he drove the tractor from dawn to dark getting the crops planted. Sometimes he was so busy with a hoe and shovel, breaking up the ground and digging irrigation ditches, that he never got any sleep at night. My Grandmother was a tither. Grandfather couldn't read so whatever she read in the Bible they did; she said tithe and so they tithed. He then would walk out to his field, take off his hat, and talk to God all over that field. God would show him what to do. He would come up with different ways and methods of farming. The Spirit of God was teaching him while he was in the field. When I helped him farm, he told me what to do out in the field and as long as I did what that old gentleman said I was almost as good a farmer as he was. All I had to do was copy him."

~ Kenneth Copeland, Founder, Kenneth Copeland Ministries
Believers Voice of Victory Broadcast
Ft. Worth, Texas

Excerpted, *"Great Lakes Believer's Convention Message"*
& Believers Voice of Victory Magazine

"Great is our Lord and of great power;
His understanding is inexhaustible and boundless.
Sing to the Lord with thanksgiving: sing praises with the harp…
Who covers the heavens with clouds, Who prepares rain for the earth, Who makes grass to grow on the mountains…He gives to the earth snow like a blanket of wool…He sends out His word and melts ice and snow;
He causes His wind to blow, and the waters to flow…
Praise the Lord!"

~Psalm 147

Paul Crouch, Sr.

"Great-Grandma Crouch was actually a preacher…Grandpa was the farmer…The story I remember most vividly was the day Grandpa Frank and his three sons, Andrew, James and John, went to town to but their first Model-T Ford…returning home Grandpa lost control crossing the Grand Fork bridge and plunged into the river below…(Great-Grandma) said, 'God healed them that same evening they did their chores!' For years the old timers around Woodward, Iowa told the miraculous story of the day God healed Frank Crouch and his boys."

~Paul F. Crouch, Sr., Founder & President, Trinity Broadcasting Network (TBN), wife Jan Crouch, Co-founder, Praise the Lord Television Program

Author of, I Had No Father But God

The Church in the Wildwood

The Little Brown Church in the Vale
Iowa

The Little Brown Church in the Vale

*"Come to the Church in the Wildwood…No other is so dear to my childhood,
Than the Little Brown Church in the vale…*
~Dr. William S. Pitts, Songwriter-1857

"On a bright afternoon in June 1857, I first set foot in Bradford, Iowa, coming by stage, and the spot where the *Little Brown Church* now stands was a setting of rare beauty. There was no church there then, but the spot was there, waiting for it," said William S. Pitts songwriter of the classic hymn, *The Church in the Wildwood*. Born in New York State in 1830, William Pitts was living in Wisconsin and visiting his fiancée Ann E. Warren in Fredericksburg, Iowa. After his stagecoach stopped on his homeward journey he suddenly saw the gently, rolling valley with native trees. Moved by the Iowa beauty and the Cedar River wooded valley, William Pitts wrote his feelings down, and of a little brown church he envisioned in a poem he named, "The Church in the Wildwood."

"Only then was I at peace with myself," he said of completing this cherished song sung by millions and recorded by artists over the years, including a memorable, heartwarming *Andy Griffith Show* episode. "What's Your Hurry," featuring the harmonious heartfelt vocals of Andy Griffith and Don Knox, perpetuating the endearing song's legacy to generations. In hindsight, William Pitts stop-over in 1857 seems profoundly prophetic for: "The hymn was written about this church before the church was ever built," says Jim Mann, The Little Brown Church's pastor and caretaker of heritage.

By the early 1860's, the local *First Congregational Church* founded as the First Congregational Ecclesiastical Society of Bradford in 1855, decided it needed a new church. After holding services in homes and businesses of church members for years they began to plan and pray for a place of worship. Incredibly, the location they chose to build their Chickasaw County, Iowa community church was the place William Pitts wrote about years before! Unrelenting through many trials and challenges the church members, many of them farmers and laborers, worked tirelessly on the church donating their time, skills, and milled lumber. Now, they were ready to paint their rectangular sanctuary and with the Civil War raging they searched for the cheapest paint to protect the wood. At the time white paint was preferred for most churches, but "the only paint to be found" was Ohio Mineral Paint, and to them "unhappily brown." Without knowledge of William Pitt, or his new poem, the congregation painted its little church brown for want of money.

By the spring of 1862, William Pitts, a teacher, with his bride, Ann E. Warren moved to Fredericksburg, Iowa to be closer to her family. In 1864, a music teaching position called him to an academy in Bradford Village, the location where his poem originated. Amazingly, this was the year the church was completed. William Pitts and his academy class visited the unfinished church, where he said, "rude seats were improvised," to accommodate them, and he was astounded that the church of his imagination was now a reality. "Pitts had written a song for a church that wasn't there. The congregation had painted their little church brown without ever hearing the song," says historic records of The Little Brown Church.

William Pitts sold the rights of his song for $25, history records for the entrance fee to Rush Medical College obtaining his degree in 1868. "I took the manuscript to Chicago where it was published by H. M. Riggins," he once said. Over time confusion arose because of the song's title and William Pitts eventually talked about this and explained his original copyrighted song, "The Church in the Wildwood," was rapidly becoming known as, "The Little Brown Church in the Vale," a line from the lyrics. In Fredericksburg, Iowa, Dr. William Pitts set up his medical practice in a white-board office building preserved today by the Chickasaw County Historical Society at, *Old Bradford Pioneer Village* and *Museum,* Bradford Village, Iowa.

Similar to William Pitts' fateful journey to Bradford, Iowa, the pastorate of Jim and Vicki Mann, seemed destined for a church encounter. "I was unaware that the Little Brown Church in the Vale even existed," says Pastor Mann. After a church member visited the Little Brown Church they returned with photographs, and while on vacation a few months later the Mann's visited the sanctuary. "We walked into the church and immediately were struck with a sense of awe." After dedicating his life to God in his late 30's, Pastor Mann says, "I had drifted away from the church during my teens and had a real need for the Lord." Later, while pasturing a church in Michigan he was first introduced to the song, "Little Brown Church in the Vale," sung in worship services. By 2005, through a series of unexpected events Pastor Mann was eventually called to serve the historic rural church.

"We have a real love for small churches and their unique ministries," says Pastor Mann. Years before, the Manns' farmed in northwest Iowa and always felt at home in the Iowa cornfields. Pastor Mann, and Vicki oversee the active ministry near the northeast Iowa community of Nashau. Celebrating its 145th year anniversary, this vintage sanctuary stands as a symbol of our agri-spiritual heritage. "On entering the sanctuary it is difficult to not pause in the doorway and bask in the sense of tranquility and peace," says Pastor Mann. Constructed in the New England Congregational style architecture, the church was designed with 110 seats, and is 50-feet high with a 10-foot square bell tower. Brass kerosene lamps are affixed to the window frames, and were donated by families at the time of construction, and no two are alike.

In the 1870's, the pews of walnut and pine were installed, and were numbered on the ends. "Our assumption is this was a time when the offering plate was not passed, and a pew tax was levied to raise funds to support the church and ministers." In the 1920's a tin ceiling was installed, and in the 1930's electric lights were added. Public attention was brought to the church in the early 20th century and credited to the Weatherwax Quartet, who performed the song around the country. Besides obtaining exposure for the song and gaining popularity, people were delighted to discover the Little Brown Church was a real church with an impressive history. Before World War II, thousands of couples wanted to be married in the Little Brown Church. During and after the war word spread of the quaint brown church, and it soon became a favorite location for nuptials up to the present day.

Amazingly, the Little Brown Church has witnessed more than 72,000 weddings performed in its history, says Pastor Mann, and each marriage is noted in carefully documented records. "The last wedding we had was number 72,918," and he estimates by the end of 2009 the 73,000[th] wedding will be performed. "The Church in the Wildwood" is played as the Bride and Groom exit the sanctuary, and sung as a response to the benediction each Sunday. William Pitts daughter, Kate Pitts Noble, composed "Our Wedding Prayer," that is requested at weddings.

Inside the sanctuary on a wall in simple, wooden frame is a hand written note from William Pitts gifting his song manuscript titled, "After 50 Years" and next to it is the conserved composition by William Pitts also in a wooden frame composed for a 50th celebration picnic. "Reports say that Dr. Pitts sang in a good strong voice," says Pastor Mann, who adds Dr. Pitts was 87 at the time and living with a son in New York. Alongside them, there is a copy of "The Church in the Wildwood" that Pastor Mann assumes is the original manuscript of William Pitts 1857 song.

Peace and tranquility, this Iowa landmark church with maroon hymn books in place on each pew and its alabaster interior with a cross on the wall behind the wooden pulpit was a church destined to exist and survived by love and prayers of its parishioners. Host to more than 40,000 visitors annually, the Mann's share their church and local history with bus tours and families from 50 states and several countries. Maintaining this undeniably cherished American icon, the Mann's say: "We feel blessed to have the entire world come here to see us."

The Church in the Wildwood

There's a church in the valley by the wildwood
No lovelier spot in the dale;
No place is so dear to my childhood
As the little brown church in the vale

Come to the church in the wildwood,
Oh, come to the church in the dale
No spot is so dear to my childhood
As the little brown church in the vale

How sweet on a clear, Sabbath morning
To list to the clear ringing bell:
Its tones so sweetly are calling
Oh, come to the church in the vale

There, close by the church in the valley
Lies one that I loved so well;
She sleeps, sweetly sleeps, 'neath the willow
Disturb not her rest in the vale

There, close by the side of that loved one
To the trees where the wild flowers bloom,
When the farewell hymn shall be chanted
I shall rest by her side in the tomb

From the church in the valley by the wildwood
When day fades away into night
I would fain from this spot of my childhood
Wing my way to the mansions of light

Come to the church in the wildwood,
Oh, come to the church in the dale
No spot is so dear to my childhood
As the little brown church in the vale

~Words & Music by W. S. Pitts, 1857

After Fifty Years

Once more I stand by the church in the wildwood;
Once more I wait at its wide open door;
Hearing the songs I loved in my childhood,
Thinking of those who have gone on before.
Here in this valley, near to the wildwood,

Bravely they wrought and nobly they won.
Now they are sleeping, quietly sleeping
Fathers and mothers, sisters and sons

Chorus: Little Brown Church, church in the wildwood,
Dearer art thou as years roll along!
Enshrined in dear hearts, loved in remembrance,
Cherished and lauded in story and song

Once more I stand by the church in the wildwood
Once more I hear its clear ringing bell,
Sending its tones o'er prairie and woodland,
Calling, "O! Come to the church in the dell."
Oh! How I love thee, church in the wildwood!
Ho! How I love thee, there's no one can tell!
Long may thy bell tones call in the faithful
Church in the wildwood, church in the dell

*~Words & Music by
W. S. Pitts, Written expressly for Picnic 1916*

R. C. Worley Prayer Chapel

Liberty University
Thomas Road Baptist Church
Lynchburg, Virginia

Jerry Falwell, Sr.

"My Great-Grandfather, Hezekiah Carey Falwell …built a large farm…(his son) my Grandfather, Charles William Falwell, built a large dairy and dirt farm… his horse-drawn wagons delivered milk from my grandfather's Meadow View Dairy and fresh vegetables from his farm to the people of Lynchburg…I've spent a lifetime walking with God, and I'm still learning valuable lessons of faith…When I was praying for a few hundred dollars, I was learning to trust God for millions. There were great victories-building Thomas Road Baptist Church and Liberty University… these taught me that God answers prayer."

~Jerry Falwell, Sr., (1932-2006)
Founder & President, Liberty University Chapel, Lynchburg, Virginia
Founder, Thomas Road Baptist Church

Falwell: An Autobiography, The Inside Story
& Building Dynamic Faith

Jentzen Franklin

"I was on a three-day fast when God called me to preach. My brother Ritchie and I started together in the ministry as evangelists. We preached as a tag team with him preaching one night and I preaching the following night. When Ritchie preached I would fast for him, and when I preached, he would fast for me. We did this for seven years while we were literally living in a cornfield in Kinley, North Carolina. God brought us from total obscurity."

~Jentzen Franklin, Senior Pastor, Free Chapel Gainesville, Georgia
Author of The Thirty Day Fast

A Farm Boy's Song: Alfred E. Brumley

In 1905, a boy was born to a Spiro, Oklahoma farm family, and was destined to write one of the most cherished songs of all history touching millions of lives. The farm boy's name was Alfred E. Brumley. One day while he was picking cotton on his family's farm, young Alfred looked up into the wide, blue sky and saw the birds flying high above the farm fields. At that moment, he was inspired. He thought of heaven and how happy he was going to be someday living for eternity with Almighty God in heaven. He began composing a song that is one of the most beloved tunes of all time. He named his song: "I'll Fly Away."

Bill Gaither

Near Alexandria, Indiana, Christian songwriter Bill Gaither grew up on his family's farm working alongside his brother Danny and parents, George and Lela Gaither. Among the many farm chores was pitching hay into the barn, harvesting corn, and milking cows in the barn. Bill revered his hard-working grandfather, Grover W. Gaither, who was a "Middle American farmer" possessed a strong work ethic. One fateful day in 1948 while milking cows, young Bill listened to gospel music playing on the "dust covered radio." Bill purchased a Stamps-Baxter Songbook with his fifty-cents milking allowance that was promoted on the radio. Later, while his father was in the barn doing chores, Bill talked to him about going to Nashville to hear Wally Fowler's All-Night Singing at the Ryman Auditorium. The Gaithers' left their Indiana farm to travel to Nashville's Ryman Auditorium for the big event, and it proved to be a trip that would change the course of Bill Gaither's life.

~ Bill Gaither, Gaither Homecoming
Christian Songwriter of the Century

Author of: Bill Gaither: It's More Than The Music, Life Lessons on Friends, Faith and What Matters Most with Ken Abraham

The Mother Church of Country Music

The Ryman Auditorium & Reverend Samuel Porter Jones

In 1885 a tent revival was conducted in Nashville, Tennessee by one of the leading evangelists of the day, Reverend Samuel Porter Jones. Attending the sacred service at the corner of Broadway and Eight Avenue was Riverboat Captain Thomas Green Ryman, considered one of the best and last of the great riverboat captains. Under the preaching of Rev. Jones, Captain Ryman gave his life to God. Known for his intelligent presentation of the Gospel, his storytelling and his wit, Rev. Jones was "a new breed of revivalist preacher," says historian Daniel E. Sutherland in *The Expansion of Everyday Life*.

After struggling with a failed law career, alcoholism and years of depression, Rev. Jones was converted to Christianity in 1872. Later, he led thousands to Christ through his ministry. His forthright messages endeared him to everyone, for example: "Precious Christ, seek these lost sheep to-night and help them to the cross. Brothers, won't you be saved? God help you to-night to decide: 'Others doing as they may, I intend to give myself to God to-night. Why wait for anything?" Obeying God by bringing the message of salvation to Nashville, Rev. Jones soon witnessed remarkable results from his labor, especially through the converted life of his new friend, Captain Ryman.

Captain Ryman decided to build a tabernacle so that Rev. Jones, and others, could hold services without charge. Inside the tabernacle, the same pews used today were made by the Indiana Church Finishing Company at a cost of $2700. Later, in May of 1890, the Union Gospel Tabernacle was christened, and completed in 1899 at a cost of $100,000. Rev. Jones conducted his first revival at the Union Gospel Tabernacle in 1894. Over the years, celebrated leaders preached inside the sanctuary, among them: Dwight L. Moody, Booker T. Washington, and General William Booth.

On December 23, 1904, Captain Ryman passed away. Rev. Jones conducted his funeral and during the service inside the packed Union Gospel Tabernacle, Rev. Jones took a vote to rename the Tabernacle to the Ryman Auditorium. A deeply touched audience responded with a standing ovation. On October 15, 1906, Rev. Jones died, and services were held at the Ryman Auditorium. In his home state of Georgia, thousands attended the memorial service honoring this man of God who devoted his life to the Gospel.

The hallowed halls of the Ryman Auditorium, *The Mother Church of Country Music*, has witnessed performances from leading artists in bluegrass, country and country gospel music throughout the decades. On the Ryman's historic stage numerous songs of faith and testimonies of God's grace have been offered. Captain Ryman's heart of giving to God's work produced a harvest of other souls touched by many of these Gospel concerts. One in particular occurred in February of 1972. Pastor Jimmy Snow, of Evangel Temple in Nashville, was hosting his opening night of the *Grand Ole Gospel Time* show at the Ryman. WSM Radio aired the performance with guests, Johnny Cash, Connie Smith and his father, country legend Hank Snow.

Pastor Snow says: "A man by the name of Happy Caldwell was there sitting in the second row that night. He was a wholesale whiskey distributor, and was in Nashville to buy a load of whiskey. He accepted Jesus that night, and went home, gave up his job and pastors a church in Little Rock, with 3 full power HD TV stations covering the state of Arkansas. He is, indeed, a true Victory of what God can do." Marking the 115th anniversary since the first Ryman revival in 1894 by Rev. Jones, the consecrated tabernacle of the Ryman Auditorium stands as Captain Ryman's gift to the world. Along with the tremendous spiritual influence of his dear friend, Rev. Jones, both continue to win souls and lead many to the good old ways, God's eternal paths.

The Ryman Auditorium
Mother Church of Country Music

Nashville, Tennessee

Part Two

...in every place your faith in God is spread abroad

I Thessalonians 1: 3

God's Work & the American Barn

"I liked to slip away into the hay barn...listening to the raindrops hit that tin roof and dream."

~Reverend Billy Graham, *Just As I Am*

On a cold, November night in the early 1820's, evangelist Charles G. Finney was in a western New York barn putting his horse up for the night. After preaching throughout the countryside, inside the barn he climbed the hayloft ladder and entered the hayloft abode where he spent "much secret time in prayer." On the musty floor, he spread an old buffalo robe and as he said, "Poured out my burdened soul to God in prayer." Brother Gale, his host, left Brother Finney alone to pray. The horse barn became a place of refuge and tranquility, and he prayed until "the burdened left me." It was a blessed place, for God was there. Charles Finney's prayers were heard and answered and the presence of God filled the barn.

Before this time, on October 12, 1821, Charles Finney had surrendered his life to God, and left a promising law practice for the ministry. One of the most influential evangelists of his day, and of all time, Charles Finney, "Dropped law and began to plead the cause of Christ," says the *Rebirth of America*. Until his death in 1872 Charles Finney continued to preach in large cities (including a New York City pastorate), and rural communities, and sometimes as he said in his autobiography, "I preached in barns."

Barns have accomplished God's work and purposes. Overtime, barns were used for church services until a church could be built. As in the case of Charles Finney's humble prayer chapel, barns made good places to get alone with God and pray. Circuit riders of all denominations preached their salvation messages in barns. Spiritually hungry communities gathered in barns. For example, in the history of *The First Presbyterian Church* of Watertown, New York, its pioneer founders met in 1803 for church services in Caleb Burnham's barn, and in the Sesquehanna River Valley of Pennsylvania, Eyer's stone barn was secured for church services in 1816. One descendant of a pioneer farm family recalled her northern Michigan neighbors gathering inside her family's barn for a circuit rider on horseback in the early 1900's. Inside the barn, Ester Forrester, a Church of God evangelist, recalled sitting in the hayloft listening to the preaching that led many to the Lord.

A barn defined in Hebrew is "a place of barley." In the story of the Book of Ruth we see how the Hebrew barley fields and harvest were important to this community. (Ruth 2:17) Barns protected the grain from spoilage, and were a valuable part of the farm. Barley was just one product they grew from the earth. "The Promised Land offered good opportunities for making a living with its water and tillable soil," say biblical scholars, J.I. Packer, Merrill C. Tenney, and William White. The story of the five barley loaves of bread given to Jesus and belonging to a young boy to feed the crowd of 5,000 was a gift of the heart. Little did this Hebrew boy's mother realize that after she had gathered the barley from the fields, and ground the grain with a stone, that the Master would use them for His purposes. God's work was carried out through the use of humble barns throughout the ages.

The first barns were rudimentary shelters built for animals and storing grains, and later improved by time, experience, and resources. Certainly no barn in history was used for God's work as the one in Bethlehem where God's only Son was born. Luke 2: 7 says, "And she gave birth to her Son, her Firstborn; and she wrapped Him in swaddling clothes and laid Him in a manger, because there was no room or place for them in the inn." Contrary to the familiar image of a wooden stable, the earliest barns in biblical times are believed to have been underground stores houses, cave-like dwellings. Although we will never know for certain exactly what type of stable Jesus was born within, scripture does reveal the subject of barns filled his life lessons.

Jesus addressed the subject of barns, and one of several is found in Luke 12: 18: "And he said, 'This is what I will do: I will pull down my barns and build larger ones, and there I will store all my grain and my goods. And I will say to my soul, "Soul, you have many years to come: take your ease, eat, drink and be merry." But God said to him, "You fool! This very night your soul is required of you; and now who will own what you have prepared…" (Interestingly, barns in this verse indicate they were upright buildings on the landscape versus the earliest underground barns.) Perhaps Jesus overheard someone make this statement while working as a carpenter; even so, his God-inspired words attracted people of all ages and cultures and eternally changed lives, then and now.

In Mark 6:3, Jesus is first called a carpenter. Most likely his craft required repairing wooden farming tools, and hewing beams for buildings, maybe barns. In early Israel carpenters made yokes for oxen and wooden plows, made of oak. Historians tell us that the earliest Hebrew plow was a forked stick with a pointed end. "Where no oxen are, the grain crib is empty, but much increase comes by the strength of the ox," says Proverbs 14:4. Oxen yokes and wooden plows were important for developing farm fields, say biblical historians Packer, Tenney and White. The grain crib, or the barn, was the hub of the farm. Farmers in Biblical days separated precious grain from the straw by having oxen trample over it. As in Jeremiah 4:3, they were also instructed to "break up your ground left uncultivated for a season, so that you many not sow among thorns."

Since the days of the Garden of Eden, God has blessed the vocation of farming. He gave definite instructions to the first farmers, Adam and Eve "to dress and keep it." In Genesis 4:2 the sons of Adam and Eve shepherded flocks of sheep and tilled the soil. Like the sacredness of boundary stones, (Deut. 27:17), God's instructions were specific and for a purpose. The Hebrew farmer depended on God and recognized that harvest blessings came from above. Difficult work, requiring dedication and commitment, especially during planting and harvest seasons. "The Israelite farmer had to deal with rocks, thorns and thistles," say historians. They feared the sun that scorched seedlings and ruined a farmer's livelihood, making the parable of Jesus in Matthew 5:45 come alive with new meaning.

The good things of the earth were blessed of God, and the seasons were respected. Their lives were centered on the farming way of life and barns were valuable to the farm, as well as silos. Unlike the cylinder-type silos standing on farms today, constructed primarily in the early 20th century, archeologists discovered underground Hebrew silos near Beersheba dating back to 4000 B.C. These underground silos were 25-feet deep and contained clay jars filled with grains of wheat, barley and lentils. In the Judean desert, a cave known as the "Cave of Letters" was also discovered in the Judean desert and unearthed several items, including a bowl filled with dry pomegranates, olives, and dates, according to historian Avraham Negev. This discovery brings to life the scripture in Haggai 2: 19: "Is the harvested grain any longer in the barn? As to the grapevine, the fig tree, the pomegranate, the olive tree-they have not yet borne. From this day on I will bless you."

"The seeds shrivel under their clods; the storehouses are desolate,
the barns broken down...."
~Joel 1:17 (NAS)

Today, nothing seems to emit a sense of permanence as the sight of an old barn on the landscape. The American barn was built for practicality and productivity. In the early years of our nation, barns were built from logs off the farmer's woodlot, and built even before the house. (If the house was built first, it is said the woman ruled the roost!) Sometimes local craftsmen were hired to build the barn for the farmer. "Early barn builders were practical and industrious…the early barn builder was a farmer who as woodsman could cut seasoned lumber from standing trees…who could quote the Bible or the Greek classics," says historian Eric Sloane. One example of this can be seen in the life of 19th century barn builder James Garfield, who as a young man constructed a barn near Orange, Ohio, and farmed while studying the Bible and languages for the ministry; eventually, he became President of the United States, James A. Garfield.

Since 1990, America has lost more than 3 million barns. Barns are falling more each year and becoming lost icons of our agrarian past. People reminisce about Grandpa's barn that is now gone, Grandma's farm that has disappeared, and the old folks' place that is now a shopping mall. Landscapes are drastically changing and broken down barns once the center of the farm are now the only reminder that there were once working farms in many areas of the country. Symbolic of some agricultural communities, barns falling in vacant farm fields reflect the spiritual emptiness and image of many rural communities without a pastor, or spiritual advisor. (See Part Five: *Village Missions of North America*)

One upright barn that imparts a message everyday is one-of-a-kind that was inspired by the life of a Christian farmer and pastor who ministered in the rural community near Howell, Michigan. In 1994, Helen Richards commissioned the mural of "The Good Shepherd" to honor her deceased husband Sherman Richards, pastor of Hidden Springs Church. Artist, Bill Kreeger, a church member, designed the pastel-hued painting affixed on the barn standing across the street from the church. "My father loved his sheep and knowing the Good Shepherd he taught his family and his church many valuable lessons of love," says daughter and current pastor Jean Richards Tulip. Once a part of a working farm, the red-gambrel barn holds many stories, "I was one of the four girls reared on the farm, and in the old barn we used to mow hay," says Pastor Tulip. Today, this generational family farm barn is still being used of God and reaching countless lives for the Gospel by its silent message.

High and wide, barns across the country showcase religious themes like: "To God Be The Glory," Illinois; "Jesus Loves You," Ohio; "The Lord Is My Shepherd" Texas. On dirt roads and in small villages signs uphold the Judeo-Christian heritage of our nation's founding: Middlebury, Indiana's "Seek the Lord," and "Jesus Is Lord of Lenawee County," Adrian, Michigan, are two among a scattering throughout the country. Barn signs and messages both patriotic and spiritual for decades have graced the sides of barns. Now, in the 21st century a wooden barn is considered too cost prohibitive to build. "You can't get boards anymore like the ones we used to get," said Virgil Wilson, a third-generation barn builder. However, a few across the country are defying statistics and some are constructed by Vermont Timber Works, Springfield, Vermont, specializing in various framing techniques and woods since 1987. One built in the early 1970's in Hillsdale County, Michigan, was the dream of Everett Wirick and called *The Gospel Barn*. On a dirt road near Hillsdale, The Gospel Barn hosts some of the most popular Gospel music groups in the nation. Director, Scott Wiley says, "My grandfather, Everett Wirick built the barn that measures 45feet by 90 feet, with framework and rafters of rough sawn oak." Presently, scores of barns are kept upright due to the efforts of farm families, community groups, and preservation organizations, and some are used for church-related ministries like those in the states of Ohio, New Jersey and California.

Today, metal barns are replacing the old wooden barns disappearing from yesterday. "They really don't make barns the way they used to, but then they don't need to," says author and agricultural photographer Grant Heilman. On the upswing across America, people are worshipping and fellowshipping in metal barns, just as in the past in wooden barns. On farms and western ranches and rodeo and cowboy events, metal barns hold services and provide an atmosphere where rural people enjoy fellowship and can be relaxed. "We are reaching people who are very agricultural, and are comfortable in this setting," says former rancher and farmer James Ballard, Director of Church Planting, East Idaho District, who establishes cowboy churches. Cowboy Churches are one of the fastest growing church outreaches in the United States, and a number of them congregate inside both wood and metal barns. God is using these barns to meet the needs of those who work close to the earth and who love farm and ranch life. Hearts are warmly touched by the love of God inside these steel buildings, and lives are permanently transformed. The Bible is preached, and some hear the salvation message for the first time.

"Honor the Lord with thy substance, and with the first fruits of all thine increase:
So that thy barns will be filled with plenty…."
~Proverbs 3: 9-10

God's work is being fulfilled through the use of barns being used to draw communities together celebrating the rural way of life. For example, the love of farming and the tradition of quilt making are combined in Ashe County, North Carolina's the "Barn Quilt Project." The Ashe County Arts Council, West Jefferson, honors the tradition of farm life with colorful hand-painted quilts on more than 50 barns. In this western Blue Ridge region, individuals and local artists paint colorful quilts and enjoyed by the public on, "Barn Quilt Tours."

Not far from this locale, a very special farm and working barn was the setting of a dedicated farm family in the piedmont near Charlotte. God was at work on the dairy farm owned by William and Morrow Coffey Graham, and Clyde Graham and their barns were filled with plenty. The Graham Dairy Farm was one of the first to imprint "Graham Farm" on their milk bottles, and it was a respected farming operation, known for their commitment to God and "no work done on Sundays."

Inside tin-roofed barns, Reverend Graham often retreated to the hayloft, especially on rainy days. "I liked to sneak away into the hay barn and lie on a sweet-smelling and slippery pile of straw, listening to the raindrops hit that tin roof and dreaming." Reverend Graham says the family dairy barn was a sanctuary that shaped his character.

One of the earliest scriptures Reverend Graham recalls is John 3: 16 that his mother taught him: "For God so loved the world that he gave his only begotten Son that whosoever should believe in Him should not die but have eternal life." When he was five years old, Reverend Graham's father, William Franklin Graham, took him to Charlotte to hear Billy Sunday, the most influential evangelist of the day. Years later this humble, North Carolina farm family's life would change after a prayer meeting in one of their farm fields by a group of praying Christian men. Reverend Graham recalls in his book, *Just As I Am*: "…That afternoon when I came back from school and went to pitch hay in the barn across the road, and one of our hired hands we heard singing. "Who are those men over there in the woods making all that noise?" he asked me. 'I guess they are some fanatics that have talked Daddy into using the place,' I replied. Years later, my father recalled a prayer that Vernon Patterson had prayed that day; that out of Charlotte the Lord would raise up someone to preach the Gospel to the ends of the earth." During those years growing up on the farm, and working inside the barns, God was indeed molding "someone," and that someone was an extraordinary farmer's son, Billy Graham, who became the answer to their prayers, and the prayers of millions.

Ruth Bell Graham

In her book, *A Time for Remembering,* Ruth Bell Graham recalled the 1950's, when she lovingly created her U-shaped log home nestled within the Blue Ridge Mountains of North Carolina. After acquiring 150-acres of land in 1954, Ruth Bell Graham went about her project to build "a roomy, old-fashioned log home to house our large family," said her son, Franklin Graham in *Rebel With A Cause*. While her husband, Reverend Billy Graham was preaching throughout the world, Mrs. Graham drove her jeep in the piedmont mountain range wearing a ball cap and army jacket. Searching for discarded treasures like old cabins for sale and ancient, seasoned native wood boards became a part of their home. Artistic, she hired local workman to build her family's Black Mountain sanctuary and in 1956, the Graham family moved into "their homestead named Little Piney Cove."

~Ruth Bell Graham (1920-2007)
Author, Poet and Missionary

*A Time for Remembering: The Ruth Bell Graham Story
by Patricia Daniel Cronwel*

Franklin Graham

In the western North Carolina Mountains, William Franklin Graham II grew up, and is the son of Reverend Billy and Ruth Bell Graham. Franklin Graham says this region in the heart of the Blue Ridge Mountains is the "…the only place I ever wanted to call home." While growing up, Franklin often camped in the mountains with his father and listened to his father's stories of their ancestors. "Daddy told me stories about Grandaddy Graham and his own boyhood days on the farm." Together, they shared their love of the outdoors and camped overlooking Black Mountain. Fall was his favorite time of year, and he often camped overnight by himself throughout his school years. President of Samaritan's Purse, Franklin Graham travels the world with relief efforts to disaster victims and "Operation Christmas Box," a ministry of shoeboxes filled with toys for children at Christmas. Today, he remains close to the land and with his wife Jane and family, live on a North Carolina farm where, as he says in, *Rebel with a Cause*, there are lots of "barn yard critters."

~Franklin Graham, President of Samaritan's Purse
Billy Graham Evangelistic Association President

Author of Rebel with a Cause: Finally Comfortable Being Graham

A Carolina Farm Boy's Prayer

In a South Carolina farm field many years ago, a lonely, young orphan boy struggled with great feelings of being unloved and unwanted. Under the hot sun, while hoeing the field suddenly he heard music like angels singing.

"I knew in that instant, without any doubt, that God loved me," said E.C. Trammell, who would become the Grandfather of one of the most inspiring authors and television ministries co-founders, Joni Lamb.

~ "Surrender All"

by Joni Lamb DayStar Network

The Old Rugged Cross Church Historic Site

Michigan

The Old Rugged Cross Hymn Historic Site:

A Former Barn, A Preserved Church
Featuring George Beverly Shea

"On a hill far away stood an old rugged cross, the emblem of suffering and shame…"

~Reverend George Bennard, Songwriter, 1913

"America's Gospel Singer," George Beverly Shea is one of the most admired Christian artists and songwriters in history singing publicly for more than 60 years. He was born on February 1, 1909 in Winchester, Ontario, Canada. While growing up in Canada, he recalled as a little boy when his mother was asked if she could play a "new" hymn—and that new hymn was, "The Old Rugged Cross." Years later, George Beverly Shea brought his profoundly beautiful rendition of, "The Old Rugged Cross," to a global audience through the Billy Graham Crusades. He has performed this timeless classic since the 1940's at the Billy Graham Crusades. Growing up, his father was a Methodist minister, and as a boy George sung in the church choir. In his poignant interview with Bill Gaither featured in the documentary, "George Beverly Shea: Then Sings My Soul," he shares how he penned the memorable music to one of the greatest Christian hymns of the 20th century: "I'd Rather Have Jesus."

Later in the late 1930's, George Beverly Shea worked as a Chicago radio announcer appearing on a popular Chicago-area program called "Songs of the Night." At that time a young pastor named Billy Graham was featured on that radio program. By 1947, George Beverly Shea gifted with a distinctive baritone voice was asked to join the Billy Graham Crusades. For decades people of all nationalities have been drawn to the "old rugged cross," and found peace in the Lord

Jesus Christ from his singing ministry. According to the *Guinness Book of World Records*, he holds the world's record for singing to an estimated 220 million people. With the love for God's Word in his heart he recorded more than 500 songs in his lifetime, and in addition featured on numerous broadcasts over his 60-year career, including The University of North Carolina Pubic Television and many other accolades, such as being inducted into the Gospel Music Hall of Fame in 1978, and a Lifetime Achievement Award from the Canadian Gospel Music Association in 2004.

"The Old Rugged Cross" is shared by many denominations and genres of artists. Cliff Barrows, highly respected song leader for more than fifty years for the Billy Graham Crusades, has said, "Music can be a mighty force, and be used by the Spirit to really prepare for the preaching of the Word." The first to record the song was Gospel singer Virginia Asher, a songstress for the Billy Sunday campaigns. Afterward, Ella Fitzgerald, Tennessee Ernie Ford, The Oak Ridge Boys, Willie Nelson, Alan Jackson, Vince Gill and Brad Paisley, to name a few who've recorded the classic hymn that songbook editors say is "the most popular of all twentieth-century religious songs."

While the beloved hymn, "The Old Rugged Cross" takes its rightful place as one of the top 10 songs in Christian musical history, the story of the *building* where the hymn's first public performance took place seems to be a shadowy footnote in the pages of American Christian heritage. Today, if not for The Old Rugged Cross Foundation (ORCF), and those who held the vision for restoration of this revered landmark our spiritual-agricultural past, the former barn would be lost in the pages of history. The story of the Old Rugged Cross Hymn site begins in 1862 when a southwest Michigan farmer built a barn in the community of Pokagon. Named for Chief Leopold Pokagon of the Potawatomi Indians, historians estimate there were 250 Potawatomi in Pokagon's group. Within this fertile agricultural center the farmer stored his hops crop inside his 28-ft. by 60-ft barn. Hops, a crop produced around this time became a booming industry especially during the Civil War. Primarily used to make alcoholic beverages, hops production dropped off in the late 1870's. From this humble beginning, the common pitched-roof barn was soon up for sale, and was sold to the First Methodist Episcopal congregation of Pokagon in 1876. Taking out a loan the congregation made renovations, adding a "two-story 30-foot section, painted, leaded glass windows, pews, and a belfry with a cast bell."

Years later, this obscure agricultural structure was destined to become the site of the first public performance of one of Christianity's most beloved hymns, "The Old Rugged Cross," was written by Reverend George Bennard.

In 1913, Rev. Leroy O. Bostwick, the pastor of the First Methodist Episcopal, held a series of revival meetings and invited his songwriter-friend, Rev. George Bennard, to assist him. Anticipation ran high as the *Dowagiac Daily* reported Rev. Bennard's approaching visit: "Mr. Bennard is a sweet singer and a splendid gospel preacher." When Rev. Bennard arrived in January 1913 he brought his unfinished song that was first begun at his home in Albion, Michigan. "I began to write 'The Old Rugged Cross,' and composed the melody first," he once said. The first verse and chorus of the hymn were written in Albion, says author George Passic.

As revival fires illuminated the hearts of the lost, and souls looked heavenward, inside the former barn's sanctuary. Rev. Bennard was powerfully moved by God's spirit, for during his stay in the quietness of the church parsonage he finished his epic hymn. Overjoyed by his completed composition he later told a friend about the experience: "I sat down and immediately was able to rewrite the stanzas of the song without so much as one word failing to fall into place." One source says he called Rev. Leroy Bostwick's wife to listen as he took out his guitar and sang her the completed song.

Gathering his guitar and his finished composition in hand, Rev. Bennard took his place inside the rustic church and publicly sang the hymn in its entirety for the first time at the Pokagon revival of 1913. While introducing "The Old Rugged Cross" to those attending, he was accompanied by church choir members Clara Virgil, Frank Virgil, Olive Marrs, William Thaldorf, organist Florence Jones and violinist Arthur Dodd. According to a Michigan Historic Marker, Rev. Leroy Bostwick financed the first printing of the hymn.

Some people may wonder what prompted Rev. Bennard to write the song. The only explanation Rev. Bennard gave regarding the song he penned at this time was that it originated from a deeply personal spiritual experience: "I seemed to have a vision . . . I saw the Christ and the cross inseparable." Born on February 4, 1873, in Youngstown, Ohio, Rev. Bennard sensed God's call on his life at early age. At 16, his father died and the young man worked as a coal miner to support his family. Later, he served as a Salvation Army officer in Illinois, and preached throughout the United States and Canada.

By 1915, "The Old Rugged Cross" was becoming a cherished hymn by millions. Rev. Bennard's timeless song was catapulted into the hearts of American's and Canadian's, mainly through the efforts of the gospel music publisher Homer Rodeheaver. Born on a farm in rural Ohio, Homer Rodeheaver was leading music at a Kansas revival when the great evangelist Billy Sunday attended the services. By 1910 he was the song director for Billy Sunday's crusades. Homer Rodeheaver purchased Rev. Bennard's song for $500, and one source says. "The Old Rugged Cross" was subsequently recorded 11 times (along with "In the Garden"), selling a million and a half copies by 1915. Billy Sunday's crusades brought tremendous exposure to the heartfelt lyrics and melody introduced in the Pokagon converted barn-church. Sadly, though the hymn was getting its rightful place in the spotlight, the church where it was first performed was fading into the shadows. Yet again, according to the Michigan Historical Marker, the church was sold to a local farmer named John Phillips, "who used it as a storage barn."

Neglected for 82-years, the historic Old Rugged Cross Church was saved from collapse through the generosity of a local couple, Bob and Molly Shaffer, who purchased it in 1998. "We felt God's call to save and reclaim His house," says Molly Shaffer, ORCF treasurer. The Shaffer's bought the 90-foot-long building from a local family. The structure is considered one of the last Civil War era buildings standing in Michigan. "In one place, the church wall was resting on the ground and the whole thing was sagging."

Now painted white, the church's restoration began in 2000 with multiple renovations conducted, including a new concrete foundation and replica bell tower in 2007. (Chapter Photograph: see endnotes)

"God has led us in this effort," says Bob Shaffer, ORCF president and restoration project manager. Now designated as one of our nation's treasures, the Church is listed in the National Register of Historic Places. The Old Rugged Cross Foundation's Board of Directors is unique with ties to the hymn and from various Pokagon churches," says Marta Dodd, whose grandfather was the violinist at the hymn's debut in 1913.

In 1998, the Pokagon United Methodist Church (PUMC), the descendant of the 1913 Methodist Episcopal congregation, celebrated the 85th anniversary of the hymn's performance. Likeminded individuals organized the event together with volunteers of all denominations and created the half-acre Old Rugged Cross Memorial Garden directly behind the historic church. A 14-foot-tall wooden cross stands on a hill near the restored church, "as a reminder of Christ's sacrifice for us." Peaceful, the garden provides benches for prayer. Throughout the garden, a brick memorial walkway holds the names of people contributing to the preservation of this historic building. In 2009, the historic church held is first wedding in more than 90 years, the wedding of a family member of former pastor, Rev. Leroy Bostwick's and "orchestrated by God," said Bob Shaffer.

Before his death in 1958, Rev. Bennard ministered all over the country, and Grass Lake, Michigan is one community that continues to honor the legacy of the Old Rugged Cross hymn and the life of Rev. Bennard. In 1938, Rev. Bennard was asked to be guest speaker by his friend Rev. Forrest Cook for services held on Sackrider Hill near Grass Lake. Named for area pioneers, Christian and Jane Sackrider, the hill's perfect location originally was selected by a local minister, Pastor Harold Solomon.

A cross was placed on the hill for Easter sunrise service in 1938, and later in 1950 a permanent cross was erected. As subsequent crosses deteriorated, new ones replaced them. The "Old Rugged Cross" hymn is honored each Easter sunrise service by the Grass Lake Ministerial Association. "On Sackrider Hill we have a reminder of that old rugged cross on a hill far away in Jerusalem," says Pastor Chuck McNeil of the Federated Church of Grass Lake, of the Grass Lake Ministerial Association, in the Grass Lakes Times, published by Alex Weddon.

Near Reed City, Michigan *The Old Rugged Cross Museum* has been caretaker of Rev. George Bennard's memorabilia for nearly 20 years in association with the Reed City Area Genealogical Society. Safeguarding many artifacts from Rev. Bennard's songwriting and ministry career, the guitar that was used for the performance of the "Old Rugged Cross" hymn in 1913 is preserved, according to the museum curator. After the death of his first wife, Rev. Bennard married Hannah Dohlstrom, in July 1944. Hannah Dolstrom Bennard was his pianist, and they settled on her family farm after their marriage, not far from Reed City on Old 131. Treasures in secret places, this museum holds the piano used for their song crusades in one room devoted to their memory. A photograph taken around this time in front of their farmhouse depicts a sign stating: "Home of Living Author, Rev. George Bennard," and standing beside it a tall wooden cross reading: "The Old Rugged Cross."

Rev. Bennard's final years were spent on this farm, and spokesperson David Bonfell of The Old Rugged Cross Museum says, "They spent six months on the farm, and six months in California." In 1954, four years before his death, Reed City erected a large cross honoring the hymn and the life of Rev. Bennard. The cross fell during a storm; as well as subsequent crosses put up over the years. Rev. Bennard died on October 10, 1958 of asthma and was buried in Inglewood Park Cemetery, Inglewood, California, according to an Inglewood Park representative. On his farm in 1958 before his death, a rare historical recording is preserved of an interview with Rev. Bennard and of him singing his now famous song, "The Old Rugged Cross." At that time, Rev. Bennard revealed more about his journey and composition, saying he wrote it over a three-month period and "helped me know a deeper meaning of the cross." Rev. Bennard emphasized his hymn was inspired from his Salvation Army years. Perhaps we will never know the full extent of Rev. Bennard's suffering or life experiences leading him to pen, "The Old Rugged Cross." But we do know he was a man of integrity and absolute devotion to God and his fellow man. A consecrated life, a humble life, Rev. Bennard's dedication to the service of the Gospel continues to be revered by several worthy organizations, and locales; each of them holding his place and his song, "The Old Rugged Cross," an endearing place in our American agri-spiritual history.

"He who goes forth bearing seed and weeping
(at needing his precious supply of grain for sowing)
shall doubtless come again with rejoicing, bringing his sheaves with him."

~Psalm 126: 6

Belarmino "Blackie" Gonzales

"God is the one who calls and appoints each of us. Those who will listen and heed His voice are the ones He will use….When you know that God has chosen you, then you will have the confidence to walk by faith, to live by faith and to walk in His love…Once I realized He had chosen me for a specific work, I had the confidence to know that He would direct my path and accomplish the work He had give me…He is the one who has brought everything in this ministry to pass, all to the glory of Him."

~Belarmino "Blackie" Gonzales (1925-2008)
God Answers Prayer Television Broadcast

SUN Broadcasting Founder, Sante Fe, New Mexico
Author of: Miracles…On La Voz De Cristo Rey: A Biography

John Hagee

Pastor John Hagee tells the remarkable story of how he once owned a farm thirty years ago, but sold the farm to establish their ministry television network, GETV. Pastor Hagee said how much he loved that farm, and declared by faith, 'Someday God is going to give us another farm!' With his farm now sown for the sake of the ministry, Pastor Hagee then drew out a dream farm on paper and began to search to locate his ideal farm. But over the course of time he began to get discouraged, unable to find the exact farm he desired. However, his wife, Dianna Hagee, kept looking, until one day their Texas farm was finally realized. "We've had our farm six-years and our farm is much bigger and better than the farm of my dreams!"

~John Hagee, Senior Pastor and Founder, Cornerstone Church,
San Antonio, Texas

Author of Life's Challenges Your Opportunities

Kenneth Hagin, Sr.

Pastor Kenneth Hagin, Sr., founder and former president of Rhema Bible College was beloved and respected Bible teacher in the body of Christ and leader of the Word of Faith movement. His rural Texas roots and spiritual depth endeared Pastor Hagin to millions around the world. The scripture text he most often preached from was Mark 11: 22-24: "And Jesus, replying, said to them, Have faith in God (constantly). Truly I say to you, whoever says to this mountain, Be lifted up and thrown into the sea! And does not doubt at all in his heart but believes that what he says will take place, it will be done form Him. For this reason I am telling you, whatever you ask for in prayer, believe, (trust and be confident) that it is granted to you, and you will get it."

Kenneth Hagin, Sr. (1917-2003)
Founder, Rhema Bible College
Tulsa, Oklahoma

Kenneth Hagin

"You and I have a responsibility if we are going to grow and increase in God's love! For example, when I was a child, my grandpa had a garden in his yard. When it rained, the sun would bake that north central Texas blackland soil until it had a hard crust on it. My grandpa would go up and down those rows of plants with his hoe loosening the soil around the plants so the hardened clay wouldn't choke the life out of the plants. If the surface of the soil was hard, then when it rained, the moisture would just run off that hard surface. But if the soil was broken up so it wasn't hard, the rain would soak into the ground so the plants could grow. Many of us are like that hardened soil. Our hearts have gotten hard! But if we want the love of God in us to produce a harvest, then we must make sure the ground of our heart is continually soft, so the love of God can grow unhindered."

~Kenneth Hagin, Senior Pastor, Rhema Bible Church
President of Rhema Bible College, Tulsa, Oklahoma

Author of, From A Pastor's Heart

*"The harvest is past, the summer has ended
and the gathering of fruit is over, yet we are not saved!"*

~Jeremiah 8:20

Marilyn Hickey

Bible teacher, evangelist and author Marilyn Hickey has traveled the globe for decades, and is a trailblazer holding ministry crusades in countries where women have never preached before. Growing up on an apple farm in Sewickley, Pennsylvania, Marilyn worked in the family orchard and sold apples at a farm stand. "I had to walk to through those orchards every day to go to school. We were able to recognize all the kinds of apples…It was really enjoyable," she writes. Canvassing the earth with the word of God as her mission, Marilyn Hickey is also co-pastor with her husband, Wally Hickey, in Denver, Colorado.

One valuable lessons derived from her years living in farm country is taken from that experience. Pruning and spraying non-productive apple trees eventually they would be cut down if they didn't bear fruit. "Jesus said that! He said, 'the axe is laid to the root of the tree.'…He would cut down the corrupt trees that were producing corrupt fruit…I want to see more fruit. I want to be more than a conqueror…here's how: 'That Jesus Christ may dwell in your hearts by faith; that ye, being rooted and grounded in love' (Eph. 3:17)…Faith works by love! …They have to go together."

~Marilyn Hickey, Founder, Marilyn Hickey Ministries
Happy Church, Denver, Colorado

"Covering The Earth With The Word" Television and Radio Ministry
How To Become More Than A Conquer

The Gezer Calendar

God has blessed the vocation of farming since the first days of the Garden of Eden. Seasons were respected and harvests were celebrated. Farming was an honored profession in ancient Israel and a vital part of their culture. From the beginning, inspired men of old wrote about farm techniques. The Greek word *agros* means, a field, a piece of ground, and where the word agriculture comes from. A sacred lifestyle, scholars believe agriculture was taught to young boys in ancient Israel. In 1908, archeologists discovered an excavated poem called *The Gezer Calendar*. Perhaps the Lord Jesus Christ recited it as a schoolboy:

The Gezer Calendar

His two months are olive harvest
His two months are planting grain
His two months are late planting
His month is hoeing up of flax
His month is harvest of barley
His month is harvest and feasting
His months are vine tending
His month is summer fruit.

Benny Hinn

"Every farmer who sows his seed into the ground expects a harvest…with his eye on the anticipated harvest, the farmer forges ahead day after day, enduring long hours of strenuous work because he understand three important principles about seedtime and harvest: He will reap what he sows, he will reap much more than he sows, and he will reap at the appointed season….your obedience to the Lord in the area of giving releases God's abundance to supply all your need according to his riches in glory by Christ Jesus (Philippians 4: 19) The key to your harvest and to God's supply rests in your seed…"

~ Benny Hinn, Founder and Pastor of Benny Hinn Ministries,
Irving, Texas

"This Is Your Day" Television Ministry
More than Enough…Living in Expectation of Supernatural Harvest

Rex Humbard

Rex Humbard was born August 13, 1919, in Little Rock, Arkansas and came from a family of preachers. "The prayers of my parents," were the first words he remembered as a child. "Evangelists, they lived by faith and taught their family to live by faith. "We had an old ragged tent for some of our meetings…One day when I was a boy, I sat on a fence in the hot Arkansas sun, watching Ringling Brothers Circus come to town…that dusty summer afternoon I watched people rush to the entrances. I clutched my fists and said to myself, "If God had a tent like that, He'd have a crowed like that!" Young Rex Humbard dreamed of the day that he could honor God, and "Put God on Main Street."

~Rex Humbard, Founder and Pastor of Cathedral (1913-2006)

"The Soul-Winning Century: The Humbard Family
Legacy 100 Years of Ministry
Excerpted from: Put God On Main Street: Rex Humbard Story

Dexter Avenue King Memorial Baptist Church

Montgomery, Alabama

Part Three

...by faith they overcame...

1 John 5:4

Our Sacred Sanctuaries:

Old Churches, Mission Bells & Cherished Landscapes

Featuring The Dexter Avenue King Memorial Baptist Church

"I accepted the call."
~Dr. Martin Luther King

The Dexter Avenue King Memorial Baptist Church of Montgomery, Alabama, is one of the most recognized and distinct churches in America and throughout the world. More than brick and mortar, the church stands as a spiritual monument, reflecting the amazing faith of its visionary leadership and faithful congregations. Existing for more than 130 years, it is only a short distance from Alabama's capital building, the place where the Confederacy of Southern States convened in 1861. Dexter Avenue King Memorial Baptist Church continues to meet spiritual needs of the community, just as when it was first founded in 1877. A peaceful place where God is honored, the sky-blue interior with dark, wooded pews is a church like none other.

    Named for the street where the church resides, Dexter Avenue, the street was originally called Market Street in the 1800s, and renamed to Dexter Avenue during the time of Reconstruction. The property where the church stands was once the site of a slave trader's pen, "where masters penned slaves to await their purchasers." Alabama Department of Archives and History records indicate Montgomery possessed four slave depots, and one of them was at Market Street (Dexter Avenue), where they say, "Slaves of all ages were auctioned, along with land and livestock, standing in line to be inspected." From this sacred ground that formerly held the sorrowful, nameless souls held in cruel bondage, a fortress of freedom arose and a sentinel of hope for all oppressed people.

Christening their church *Second Colored Baptist Church*, the church trustees purchased this Market Street land, along with a small building already on the site, for $270 on January 30, 1879. The first pastor of the Second Colored Baptist Church was Charles Octavius Boothe. Only after a few years released from the bonds of slavery, the church was a beacon of light to those who were now free men, such as Pastor Boothe, whom church history says was a freedman, "an author and scholar." Some of the parishioners were most likely connected to agriculture in some way, whether by force or by choice. Montgomery County was known as "one of the richest agricultural districts in the south," according to *The Alabama State Journal*. Although not a church member to the author's knowledge, a Montgomery ex-slave spoke to a WPA project writer in 1939 about his years in slavery. Allan Turpin, a seventy-one year old ex-preacher and farmer said, "The happiest and best life a man is ever to know in this world waits on a farm." Various occupations and trades were represented among the congregants at The Second Colored Baptist Church, and it grew and prospered. By 1883 construction was underway for a new church. Church member and builder, William Watkins, was the contractor. The designer was Pelham J. Anderson, who created a church that would forever be among the most cherished churches ever built on the southern landscape. At the time the church plans were going forward, Dexter Street was undergoing renovations, and red-clay bricks that were tossed aside as unusable were an answer to prayer to the congregants. They hauled the red bricks away in their "dray wagons to the site to be used in laying its foundation." Overcoming faith prevailed, and strategies to "hinder and suppress" did not succeed. By 1885 the first worship service was held in the basement of the renamed, *Dexter Avenue Baptist Church*. Staying true to its heavenly vision and faithful to God's direction, blessings soon followed. On Thanksgiving Day of 1889 the first worship service was held in the upstairs sanctuary of the stately Victorian style church. As the Bible was read and prayers were offered, little did these giants in the Christian faith realize the spectacular feat they'd accomplished, and that God would use their sanctuary generations later for the good of all mankind and as a church celebrated by for decades by millions.

Today Reverend Michael F. Thurman oversees the ministry, and is the twenty-seventh pastor to serve during its history that includes Dr. Martin Luther King, Jr., who accepted the call by the church in 1954. In his autobiography Dr. King recalled that momentous time in his life, and the night before he delivered his first message, reminding himself: "Keep Martin Luther King in the background and God in the foreground, and everything will be all right." Dr. Martin Luther King Jr., served as pastor until 1960, along with his wife, Coretta Scott King and family, Yolanda, Martin Luther III, Dexter Scott and Bernice. The National Park Service designated the church a national historic landmark in 1974. Later, in 1978, the church was again renamed to honor the memory of Dr. King, officially declaring it to be *Dexter Avenue King Memorial Baptist Church*.

Throughout America there are unique churches that are treasures to our culture, our spiritual foundation, and that hold historical significance nationally. A number of them, like the Dexter Avenue King Memorial Baptist Church, continue to be maintained and used for worship services. Another example is *Second Baptist Church of Detroit,* Detroit, Michigan, one of the oldest African-

American churches in the nation, established with a "mission to free the slaves." From 1836-1865, the church "received 5,000 slaves" on their way to Canada. In March 1836, after permission was granted from the Michigan Legislature, thirteen "determined men and women" established this church, considered the oldest church for "Blacks in the Midwest." Since its start, the church has been functioning for more than 170 years in downtown Detroit. An active congregation and its twenty-third pastor, Dr. Kevin M. Turman, who serves as a Chaplain with the United States Naval Reserve with a rank of Commander, maintain the priceless heritage.

Scores of our sacred sanctuaries, too numerous to mention, are maintained and preserved by descendants of the original members and are worthy of recording, even now a vital part of their communities. One example is *Walnut Hill Presbyterian Church* near Lexington, Kentucky, recognized as the oldest Presbyterian Church in the state. Built in 1801 during the time of the Cane Ridge Revival (Part One), its heritage began when the grandfather of Mary Todd Lincoln, wife of President Abraham Lincoln, donated the land the brick building stands on today. Now after 280-years, Walnut Hill Presbyterian Church is maintained by a devoted congregation who reaches out to their community and are caretakers of its unmatched heritage.

A cherished log church landmark near Johnson City, Tennessee, is surrounded by the picturesque Appalachian Mountains and occupies the distinct position of being the oldest church of any faith still standing in its original location in Tennessee. *Sinking Creek Baptist Church* was founded in 1778, when Virginia evangelists and brothers, Charles and John Chastain led a revival, and the building of the Old Log Church began for the congregation converted from their ministry. Since 1961, safeguarding this 18th century primitive sanctuary and its spiritual legacy is Pastor Reece Harris. In 1962, Sinking Creek Baptist built a new church with "the old log church still visible," and continues to impact the same region where its founders labored more than 235 years ago.

"Whatever is born of God is victorious over the world; and this is the victory that overcomes the world, even our faith," the Bible says. Faith prevailed as pioneers of tremendous conviction and courage journeyed into the unknown wilderness to establish farms and build their churches. Some were persecuted for their faith and culture. On the outskirts of *St. Lorenz Lutheran Church* of Frankenmuth, Michigan, farms with large red barns are scattered across the flat landscape of the Saginaw Valley. Founding families, and their descendants, coming from Bavaria, Germany, established most of the farms in the 19th century. Historic accounts say before launching their sailing vessels to the United States some sang in unison, "All for the Glory of God." Stirring tales of arrival in 1846 tell of some being "pelted with small stones," and scoffed at their unusual dress and native language, says Herman F. Zehnder in *Teach My People The Truth*. With a firm belief in their vision for the mission colony called Frankenmuth (derived from the words *Franconia*, or homeland, and *Muth* meaning courage) they settled near the Cass River. Today, above the white steeple of St. Lorenz Church rises above the buildings of this historic town known for its Christian heritage and Frankenmuth Bavarian Inn founded by agricultural preservationist and farmer, William Tiny Zehnder.

"Wherever you see abandoned country churches you will usually find abandoned farms, too," wrote author and historian Eric Sloane. This fact can be seen today more than ever in North Dakota's rural areas. Wide, open crop less fields once were the primary source of revenue for these communities. In North Dakota, many nationalities settled building their spiritual outposts in close proximity to their farming communities. "Every six miles, a day's wagons ride apart," these white-steeples represent the lives that built the churches and worshipped inside. With the population shifts and dwindling farms, many are in decay and disappearing. Today, a number of rural churches are lovingly preserved by local congregations and grassroots organizations like *Prairie Churches of North Dakota*. Officially a project of the national *Save America's Treasures,* Prairie Churches of North Dakota goal is to save "precious prairie places" before their demise. North Dakota has more churches per capita than any other state. On the prairie and small farming communities in 1999, out of 2,200 churches documented 500 churches abandoned. *Prairie Churches,*™ are generating enthusiast who want to protect these sanctuaries of faith. Historians and preservationists are teaming up to secure them before they disappear. A few are being turned into community centers and museums. "In some ways, they are the most important buildings in the state," says Preservation North Dakota's former executive director, Dale Bailey.

In northeast Florida on the campus of *Peniel Baptist Church* in Palatka, a history-making sanctuary where Reverend Billy Graham was ordained to the ministry in 1939 still stands. Known as the *Old Peniel Baptist Church* it remains in the same location, however, it is no longer used for regular services. Today, it is referred to as the campus chapel, and used for children's meetings. The chapel also holds some of the congregation's memorabilia from its founding in 1852. Close to the chapel at Silver Lake, Pastor Cecil Underwood baptized Reverend Graham in 1938. Today, the new Peniel Baptist Church built in 1962 is located on the campus grounds near the Old Peniel Baptist chapel. In the 1980's, the Old Peniel Baptist chapel was the church home to *All Saints Anglican* of Palatka, Florida. "After meeting in a woman's club and a funeral home, the parishioners were looking for a home to worship, and offered the Old Peniel Baptist by a community member," says Reverend John R. Jacobs, rector for ten years. "We rented it for several years before building our sanctuary," Reverend Jacobs says, adding among minor repairs they installed an air-conditioner. All Saints Anglican Church held their first service at Old Peniel Baptist on June 27, 1980, and used the church from 1980-1987. Purchasing five acres of land for a new church, All Saints Anglican Church constructed their sanctuary holding their first services on December 20, 1987.

"A major American force, as stubborn as its white spire against the changing landscape," our sacred sanctuaries sheltered the hopes, dreams and prayers eternal of the faithful. White spired steeples pointing upward toward heavens still continue to remind us our finiteness and an ever-present, unchanging God. Near Philadelphia, above the tree-lined panorama of rural Trappe, Pennsylvania, with its white steeple sitting securely on the rooftop, stands *Augustus Lutheran Church* built in 1852. Nearby, on the sloping green landscape the oldest unaltered Lutheran church building in continuous use in the United States sits and takes its place in history. *The Old Trappe Church* was built in 1743 and is of German-Gothic architecture. Earlier, when the "Father of

American Lutheranism," Henry Melchior Muhlenberg, arrived from Germany to minister to immigrant families in 1741, his first message was first "preached in a barn." Like stepping back in time, The Old Trappe Church is pristinely preserved with the original oak wood pews and doors still intact. Out of the nearby creek, stones were brought by sled for the foundation, and stonemasons marked their handiwork with signature markings, some still seen today. Constructed with two floors, the church seats around 200 and the second floor balcony rises 20 feet from the main floor. "The church has handcrafted door latches and the oak for the church was cut, hewed and prepared in the area for building in the spring time. The cast iron door latches was forged by area blacksmiths," says curator Martin Gotsch, a church member for 30 years, adding, "Its design is typical of church building for the time in northern Germany." Now after nearly 270 years, the Word of God continues to be preached just as it was when early settlers first entered into the Pennsylvania wilderness. "People kept warm in the balcony with hay. There was no heat in the building," says Martin Gotsch, who adds that without electricity in the church, celebrations like Christmas Eve are conducted by candlelight. "People got tired of pumping the old organ, and so we installed electric cords for an organ," says Gotsch, curator for two years and a volunteer at nearby Valley Forge Park. Acoustically sound, a microphone is not needed. "You can even hear a whisper."

The Old Trappe Church is used in the summer months, where the Lord's Table is practiced along with worship as it was nearly 300 years ago. It's open to the public by appointment only, as it takes several narrators to share the history of the venerated building. A cherished pre-Revolutionary War church, its cemetery dates back to the 1600s, and there are sixteen Revolutionary War veterans are buried there. Speaking volumes to us today of the fortitude and faith of these early settlers, and generations of caregivers, Martin Gotsch sums up his feelings of Old Trappe Church: "When I read aloud the scriptures as a lectern, it's thrilling to be apart of something I love dearly, and its history. It's a great place to worship God."

Churches were built near the center of small towns and villages, like the New England village scene in the front of this book, and were the hub of community life for generations. Endearing aspects of country churches were their bell towers that rang out the call to worship on the Sabbath, also called First Day. Yesteryear, traveling by horse, buggy, farm wagon, or on foot, the lives of pioneers and their descendants were ordered by the sounds of the bells. They brought a rhythm of life that has been erased from the landscape today. The sound of church bells proclaimed emergencies in the community, such as a barn burning or a range fire, and summoned the community to action. Although some still perform the tradition, most church bells are silent.

In Montana's Bitterroot Valley the western outpost of Historic St. Mary's Mission stands near the 9300-foot elevation of St. Mary's Peak near Stevensville. Founded in 1841 by Father Pierre DeSmet, the mission was established to serve the native Salish Indians "when Montana began." In 1866 Father Ravalli built the mission chapel that has been restored. Father Ravalli used earthbound material for the chapel's interior; "vermillion clay for the reds, blue from indigo traded among the tribes, (and) yellow, a sacred color, came from a cave at the mouth of the Judith River."

After the native Salish People (Flat Head Indians) were removed to the Jacko Reservation, the mission was closed in 1891. In 1954 the chapel last served as a parish church. A spiritual landmark with an agrarian past, the chapel bell of Historic St. Mary's Mission rings twice a year, in April when it opens to the public and at the season closing in October as it did more than 140 years ago. Today it continues to be an agricultural-based community. Director of Historic St. Mary's Mission Colleen Meyer, and husband, manage a cattle and hay ranch and have been in agriculture all their lives. Today, the native First Peoples culture and heritage is honored and preserved through the efforts of Historic St. Mary's Mission, "This is a very unique place where the Indians welcomed the Black Robes," says director Colleen Meyer.

Near Fairplay, Colorado, two churches from the gold mining-era are a part of the South Park City Museum and Historical Foundation. South Park City Museum is located in Park County's grassland basin and holds the Sheldon Jackson Memorial Chapel built in 1872 by Dr. Sheldon Jackson, and the Dyer Memorial Chapel, both preserved by the museum. The Presbyterian Historical Society lists the Sheldon Jackson Chapel in its Synod of Rocky Mountains Records as Site No. 259. As a Presbyterian pastor and pioneer missionary, Dr. Sheldon Jackson chapel is preserved close to historic ranches representing the 100-year history of agriculture in the area.

The Dyer Memorial Chapel is a log structure that once was the property of John Lewis Dyer, a Methodist circuit rider and one of the most beloved itinerant preachers in Colorado history. Prior to surrendering his life to the ministry, John L. Dyer was a farmer and miner in Wisconsin and Minnesota. After arriving in Colorado in 1861, "Father Dyer" traveled by horseback throughout the rugged mountain terrain and boldly shared his faith. Known as the "Snowshoe Itinerant," he purchased the building for $100.00 from the mining settlement of Montgomery and dismantled the building numbering each log and moved it to Fairplay in 1867. "Providing spiritual guidance in the early mining camps," John Lewis Dyer's illustrious exploits bringing the Gospel to miners was said to total 10,000 miles throughout his ministerial career. The year before his death in 1901, John Lewis Dyer's pioneer life was commemorated by the State of Colorado, recognizing him as one of the sixteen founders of this frontier state. In the state capitol dome a stained glass image honors the life of the "Snowshoe Itinerant." Close to the city of Leadville, North America's highest incorporated city at an elevation of 10,000 feet, pioneer board houses are still visible from early settlers.

America's adobe churches hold unique places within America's agri-spiritual heritage. Adobe missions were some of the first church bells to ring out across the continent. Adobe missions were for the most part made from red clay, and located mainly in the south and southwest regions of the nation. The portion of adobe missions featured honor their heritage, and pay tribute to first peoples who built and found solace within them: After 450 years the legacy of the Cathedral-Basilica of St. Augustine, St. Augustine, Florida founded in 1565, is the oldest continuously active Catholic faith community in the continental United States. The existing St. Augustine Church was constructed and completed in 1797, and is listed as one of the oldest churches in the United States. Later, in the 19th century, a Spanish Renaissance bell tower was added and called worshippers of

St. Augustine, "America's Oldest City." America's oldest church resides in the southwest region of Santa Fe, New Mexico, the nation's oldest capital, founded in 1610 and called San Miguel Mission. Under indigo skies and in the midst of rich Spanish culture San Miguel Mission stands in the same location and built in 1610-1626. The actual mission was founded earlier in 1598. A treasure on the agri-spiritual landscape it is still intact after nearly 400 years. Native Indians constructed the Pueblo style mission that contains priceless wooden altar screens dating back to the 1700's. The U.S. Department of Interior's Historic American Builder Survey in 1934 examined San Miguel Mission, cataloging its heritage and among the rare attributes was a bronze bell called the San Jose Bell. Santuario de Guadalupe was built in 1781 by Franciscan missionaries and is preserved at 100 Guadalupe Street in Santa Fe, and holds artifacts from mission days. One artwork is of the Spanish Southwest by Mexican artist Jose de Alzibar dated 1783, and history records it was painted expressly for this church and brought by a mule caravan.

Arizona's oldest Spanish mission, *San Jose Tumacacori,* stands south of Tucson in Tumacacori National Historical Park. Federally protected since 1908, it was originally established in 1691 by Jesuit missionary, Father Kino. The adobe church standing on the mission grounds today was later constructed in the early 1800's and deserted in the 1840's. Other old missions located nearby are: Los Santos Angeles de Guevavi built in 1751, and San Cayetano de Calabazas, established in 1756, two preserved adobe ruins that represent the spiritual and agrarian landscape of this ancient place. In the Santa Cruz Valley of Tucson, Arizona, the "Dove of the Desert," *San Xavier del Bac Mission* stands as a white sentinel reminding us of the missionary efforts of the "black robes" and built between 1783-1797. Since 1864, the mission has provided education for the Tohono O'odham families, but is "open to all."

San Juan Capistrano Mission located in San Juan, California is widely known for the legend of the swallows returning each year. Today the ancient bells that once were part of the mission are now hung on a bell wall. They are rung only for special occasions, like Swallows Day, says spokesperson Patricia Margosian Terrell. "Father John O'Sullivan started the legend and celebration of the Cliff Sparrows in the early 20th century. Swallows do frequent Southern California, and therefore San Juan Capistrano." Serra's Chapel built in 1782 is unrivaled as the oldest in California still in use today. Known as the "Jewel of the Missions," its history is documented and reports at one point a thriving agricultural community; growing grapes, pomegranates, and other fruits, as well as vegetables and 8,000 head of cattle, 4,000 sheep…maize and beans.

Not forgotten, these irreplaceable treasures represent the spirit and soul of those gone on before us. Through the dedication and commitment of those caring for them we will continue to hear the sounds of the bells and see their white-capped spires throughout the countryside well into the 21st century.

*"The earth has yielded its harvest...
God, even our God, will bless us."*

~Psalm 67:6

T. D. Jakes

In a rural West Virginia mountain community, Thomas Dexter (T.D.) Jakes grew up learning to trust in God's promises at an early age. T. D. Jakes, recognized as "America's Bishop," as a young boy in his neighborhood was known for his strong faith in God. Called Bible Boy because he carried his Bible to school, T. D. Jakes' passion was delivering sermons to imaginary congregations. As a young man, T. D. helped his mother, Odith Jakes, in the family's vegetable garden, and after the harvest he sold the produce. T. D. Jakes learned from his parents to believe in himself. Today, a man of vision and great faith inspiring millions around the globe, Bishop Jakes encourages us all by saying: "Keep the faith. Practice a relentless belief in God, in his promise, and in the fact that He will perform His promise."

~T.D. Jakes, Senior Pastor & Founder, The Potter's House, Houston, Texas

Author of, Mama Made the Difference

Coretta Scott King

Coretta Scott King was born on a farm in rural Heilberger, Alabama in 1927, and was reared in a Christian home. Obadiah and Bernice Scott, her parents, taught her to love her fellow man even when facing oppression and unfairness. Life was very hard for young Coretta Scott, working in cotton fields during the Great Depression. She and her family endured many injustices. Later, her brilliant mind and talents took her to the New England Conservatory of Music. In her book *First Lady of Civil Rights*, Coretta Scott King records her memories of meeting and later marrying, Dr. Martin Luther King, Jr. She shares their extraordinary life and Martin's faith and calling from God for civil rights. "When God calls you to a great task, He provides you with the strength to accomplish what he has called you to do," she said. Coretta Scott King noted Dr. King's extraordinary words of faith, vision, and purpose: "He said with this faith, we will be able to hew out of the mountain of despair the stone of hope with this faith we will transform the jangling discord of our nation into a symphony of brotherhood."

~Coretta Scott King, wife of Dr. Martin Luther King
(1930-2006)
Civil Rights Pioneer & Author

Excerpted from, First Lady of Civil Rights

Dr. Martin Luther King, Jr.

Dr. Martin Luther King, Jr., was born January 15, 1929 to godly, Christian parents, Martin Luther King, Sr. and Alberta Williams King of Atlanta, Georgia. Martin Luther King, Sr., his father, was pastor of Atlanta's Ebenezer Baptist Church, and profoundly influenced young Martin. His father shared life stories of growing up as a sharecropper son near Stockbridge, Georgia. One in particular was when Martin, Sr. was a young man, he keenly detected a plantation boss was "cheating his father out of some hard earned money," and received abusive treatment for trying to correct the error. Martin, Sr., left the farm saying years later, "I ain't going to plough a mule anymore." From his mother, Alberta Williams King, Martin Luther King, Jr, was taught to value his personhood "a sense of somebodiness."

Both parents instructed him "not to hate the white man but that it was my Christian duty to love him." Bearing the rewarding ancestral linage of preachers, Martin Luther King, Jr. answered his own call from God and sensed he was to give his life to "God, who is the same yesterday, today and tomorrow." The Bible, prayer and a life of faith was the invaluable family legacy he carried with him as he stood for the first time in the pulpit, and gave his first sermon at Dexter Avenue Baptist Church in Montgomery, Alabama in 1954 saying, "I come to you with only the claim of being a servant of Christ, and a feeling of dependence on his grace for my leadership."

~Martin Luther King, Jr., (1929-1968)
Father of the Civil Rights Movement

The Autobiography of Martin Luther King, Jr

Rosa Parks

Rosa Parks was a woman of great humility, gentleness, and a devout faith in God. Born in 1913, young Rosa was reared in rural Pine Level, Alabama. As a young child she was baptized in the Mount Zion AME Church, and later was an active member and Sunday school teacher. While living in Pine Level as a child, Rosa worked in the cotton fields picking cotton barefooted in the burning hot sand of a wealthy planter for fifty cents a day. Rosa's immoveable faith and quiet trust in God led her through many difficult experiences. Her vision and strength was renewed like the eagle as she read her Bible, and her spirit soared above the merciless, but commonplace discrimination. Years of enduring hardships and injustice, author and historian Douglas Brinkley says in *Rosa Parks: A Life,* that for Rosa reading Psalms 23 and 27 gave her "the strength to love her enemy." A history maker in 1955, with strong confidence in God's promises, Rosa "refused to surrender her seat for a white passenger." Later, Rosa Parks moved to Detroit, Michigan working as a seamstress, and was a faithful follower of the Lord all her life. "Faith in God was never the question for Rosa Parks; it was the answer."

~Rosa Parks, (1920-2003)
The Mother of the Civil Rights Movement

Rosa Parks: A Life
by Douglas Brinkley

Cathedral St. Augustine
America's First Parish-1565
St. Augustine, Florida

San Xavier Mission
"Dove of the Desert"
Tucson, Arizona

Part Four

"I have fought the good fight, I have finished the race,

I have kept the faith."

II Timothy 4:7

Faith of A Quaker Miller:

The Greatest Biblical Wheat Experiment in the World

*"It was wheat-planting time in Michigan...
I was led to do something that has since been heard of all the way around the world."*

~Perry Hayden, Dynamic Kernel Experiment

On a quiet Sunday morning in September of 1941 in the town of Tecumseh, Michigan, Quaker businessman Perry Hayden was stirred by a message delivered at his Tecumseh Friends Church on the scriptural text John: 12:24: "Unless a corn of wheat fall into the ground and die, it abideth alone, but if it die, it bringeth forth much fruit." Third-generation flour miller and owner of Hayden Flour Mill, Perry Hayden said, "It gave me an inspiration to do something I had wanted to do for years...to plant some wheat and see just what would happen." For years a committed tither, he was inspired to tithe from the project saying, "I believe if we tithe each year we may be surprised."

Humble beginnings, a plot of ground measuring 4 feet by 4 feet, was spaded into 12 rows by 12 boys, and one of them was Perry's son, Billy Hayden. The first planting took place in 1940, and soon was officially named *Dynamic Kernels*. With 360 "sturdy Bald Rock wheat seeds" given by a Lenawee County farmer, the first planting and early steps of this epic test were bathed in prayer by Perry Hayden and his family, wife Elizabeth Comfort Hayden, and children, Mary Jane, Bill, Martha, Betty, and Joe. During the years until the final harvest of 72,150 bushels of wheat in 1946, the efforts of industrialist Henry Ford, and the cooperation of 276 farmers in several states, became known around the world.

Perry Hayden's initial goal was to see how much *one kernel of wheat* could produce over time. A devoted tither, his main objective was to tithe from the produce of each harvest. "When he heard Clifton Robinson preach from this text, John 12:24, the words had a peculiar resonance in

Dad's soul," says his son, John Hayden, Pastor of Columbia Baptist Church, Mason, Michigan. Not without tremendous challenges, the first crop was successfully harvested, cut with a hand sickle on August 18, 1941. Family members harvested the 18,000 kernels and carefully separated them from the chaff. "This first harvest produced 50-times the original amount of wheat," said Perry Hayden, "enough to almost fill a quart fruit jar." Out of the 50-cubic inches of wheat from the first harvest, 5-cubic inches were tithed to his local Friends Church. Dynamic Kernels would bring glory to God, and in meticulously written journals, Perry Hayden planned for the second harvest. While juggling his responsibilities at Hayden Mills, records kept track of wheat production on surrounding farms and on the agricultural holdings of Henry Ford and Ford Farms. In 1940, Ford Farms acres produced so much wheat it took 33 truckloads to Hayden Mills. During this time, Detroit automobile manufacturer Henry Ford discovered Dynamic Kernels from a national publication. Enthusiastic and supportive, he soon became personally involved. Henry Ford grew up on a Detroit-area farm and was a farm boy at heart all his life. Born in 1863, he invested much of his life and resources into his love of agriculture, resulting in numerous innovations. Besides his many conversations with Perry Hayden, he also encouraged the experiment by loaning antique farm vehicles for harvests from his *Greenfield Village and Henry Ford Museum*. Henry Ford, then in his 80s, participated in the Dynamic Kernels and enjoyed being part of the old-time farming techniques each harvest, he'd not seen since days on the farm.

Dynamic Kernels took on a new dimension after Henry Ford's presence and contributions. For example, during one harvest celebration of early farming techniques, he cut the wheat with a scythe and delighted in sowing seeds in the field by hand at the next planting. At the threshing of the fourth harvest, Henry Ford climbed on a Westinghouse Company upright steam engine built in 1882. He blew the whistle while farmers flailed the grain using flails from the Ford's Edison Institute. Perry Hayden kept the vision fresh in the minds and hearts of people as he managed each harvest celebration. The Hayden children used their gifts and talents, and recall that prayer was always the center of the home. "I can remember our family devotions and prayers that were confirmed, that's how things went all the time," says Martha Hayden Woodward, who was a young girl at the time and carries a scar from one time she scythed wheat. Mary Jane (Hayden) Wells, the eldest child and daughter, was a teen when she answered the phone one afternoon and was surprised to discover that the caller was Henry Ford. Courage arose, inviting Henry Ford and his family for lunch at the Hayden residence. Today, she recollects the afternoon Henry Ford and his family around their family table, "This experiment is going to go around the world," he said.

Often called the Biblical Wheat Experiment at the time, God was indeed at work in many ways, and goodwill was felt by all filling the Lenawee County atmosphere. At one point during a harvest celebration, Henry Ford took out a Bible and read a portion of scripture. The tithing aspect of the project drew many Christian industrialists, and spiritual zeal increased as prominent manufacturers of the day were drawn to Perry Hayden's vision. Among them: R. G. LeTourneau, (world's largest manufacturer of road and grading machinery), Harvey Fruehauf (Fruehauf Trailer Company), James L. Kraft (Kraft Cheese Company) and Kenneth S. Keyes (one of the most successful Miami, Florida realtors). Fever-pitched excitement skyrocketed as people all around the nation, and world, waited to see what God was going to do next in this southeastern Michigan farming community.

However, suddenly in 1945 an unexpected and disappointing blow struck in the form of a letter to Perry Hayden. He learned that Henry Ford was ill and would no longer be able to participate. "It left me for a moment stunned," he once wrote. Now, without the use of Ford Farms' tractors and trucks and land for the fifth harvest, everything seemed thwarted. "Reporters gathered outside our house to document this worldwide story, and were asking Daddy what was going to happen," recalls Martha Hayden Woodworth. Throughout the years, Perry Hayden had sacrificed for his God-given biblical plan. Now, bewildered and disheartened, he struggled to find a way to harvest the fifth year's wheat crop before it was ruined in farm fields covering 5,500 acres. Perry Hayden estimated that it would cost $113,000 to bring in the harvest. It would be a huge financial burden for a family-owned flourmill business.

"Henry Ford's last public activity was to help Perry Hayden stage his phenomenal six-year Biblical Wheat experiment," says chronicler, Raymond Jeffries, author of *God is my Landlord*. Now that Ford Farms was out of the picture, phone calls to the Hayden home streamed in asking him what he was going to do next and it seemed a humiliating defeat. When fear and faith collided, Perry Hayden and his devoted wife Elizabeth took their concern to the Lord. After much prayer Perry Hayden decided to go forward by faith. He set a date for the final harvest and planned for a "modified celebration." When no hope could be found, prayers were answered.

Unknown to this day why or who spearheaded the next event, in the middle of the night the sound of a brigade of Ford Farm tractors awakened the townspeople of Tecumseh. Heading toward the farm fields, author William Engle says, "Ford Farm's Allis Chalmers All-Crop Harvesters cut and threshed the wheat in one operation. With 12 trucks hauling the grain to bins on the Ford Farm and 2 gasoline trucks, 2 water tanks, and 67 men to keep the combines going, the cutting was complete." Without explanation, the Dynamic Kernels was given another opportunity to go forward and reach its final destiny. New challenges presented themselves, because after this the wheat needed to be stored and threshed.

Adrian College President, Dr. Samuel Harrison, helped at a pivotal moment as did the Tecumseh Grange and Michigan Grange, both offering to thresh the wheat, and transporting the fifth harvest's 5,000-bushels of wheat to a regional farm. Although relieved of some stress, still pressure mounted as Perry Hayden sought the Lord for the land for the final wheat crop. But where would he get the land? He ignored the comments and slights around him, and those who mocked his faith. "He was a humble man," says John Hayden. Keeping a smile on his face, Perry Hayden looked to God for answers.

Soon he learned that Dr. Samuel Harrison was spreading the word through ministries to inform farmers for the sixth planting to succeed. Why not ask hundreds of farmers to take some seed and plant the wheat at their farms? The idea grew to new proportions each day, and not long after, Perry Hayden saw a wonderful thing happening. Instead of depending on one man, farmers of all faiths and regions responded heartily saying, "Yes, we'll plant your wheat!." "At final count, there was a total of 276 farmers contributing. Like a whirlwind, the strategy was announced by the Quaker Church and agricultural periodicals. The 276 farmers of all denominations and cultures signed a note of promise to Perry Hayden to tithe their crops to their own local churches.

August 1, 1946 was declared "Dynamic Kernels Day." The Michigan Press Association offered to send free publicity, and the American Friends Service Committee extended their influence. A parade of vehicles and farming machinery from farmers all over the Midwest drove through the center of Adrian to the Fairgrounds for the sixth and final harvest. While standing on the platform of the Adrian Fairgrounds Perry Hayden quoted Romans 8:28 saying, "All things work together for good to them that love God." Mixture of emotions surfaced with the realization that the hearts and hands of 276 farmers planted the wheat seed on 2,666 farm acres throughout the nation. Hauling bags of wheat seed on cars, trucks, and tractors, an audience of more than 10,000 watched as the 72,150 bushels of collected wheat was ceremoniously ground by two old stone mills. "It was fun to be in the grandstand while wheat was being brought in and listening to the radio announcer doing a live broadcast about the mountain of wheat that was being built," one former generational farmer recalls.

"Wheat has been man's staff of life through the centuries," said Captain Eddie Rickenbacker, Dynamic Kernel supporter and WW I Ace. After the wheat was bagged, a portion was set aside and loaded into a Wright-Patterson Air-Force Base helicopter and airlifted to an Ohio baking company. Within hours boxes of graham crackers returned, and crackers made from the wheat ground at the fairgrounds was handed to the crowed. Including dignitaries, such as William Danforth, President of Ralston Purina Company. William Danforth was given the remaining wheat flour for cereal processing and distribution of cereal to war-torn Europe. At the end of the day, Perry Hayden's dream produced a far more bountiful harvest than ever expected. Consecrated to the Lord and depending on his grace, Perry Hayden's Dynamic Kernels proved to the world that his original God-inspired idea was a worthy undertaking. It also answered internal questions about how much one-grain of wheat could produce. Statistically, if the experiment kept going Perry Hayden personally calculated an astronomical figure of the wheat produced and the land mass needed to accomplish this feat. Far beyond his imagination, Perry Hayden already achieved several world-changing records. *Ripley's Believe it or Not* and other international media published his unparalleled project. Christian fellowship among farmers of all faiths "bringing them closer together" as one farmer said, was yet another result of the Dynamic Kernels. Spiritually uplifted, men in foxholes during World War II were deeply affected by the show of unity within rural America during a time of world conflict. One chaplain in the Philippines used the Dynamic Kernels message to 700 "boys awaiting orders to go into action… and which brought 60 of them forward," powerfully filled with God, they said.

After the experiment ended, Perry Hayden gave inspirational talks to high school and Sunday school groups and traveled to Mexico in a car carrying a sign that read, "Christ First." The Hayden's shared the Gospel with many, including gatherings of the Salvation Army, YMCA and church audiences. Before his death in 1951, journalists and broadcasters referred to Perry Hayden's Dynamic Kernels as the greatest biblical wheat experiment in the world. Today, his legacy can be best seen within his close-knit family, many dedicating their lives to the Christian ministry. As a young boy growing up in Tecumseh, Perry Hayden longed to be a successful businessman to provide funds to take the Gospel into Asia. One grandson, Phillip Hayden, who died suddenly in 1991 of the same disease his grandfather passed away from, became a missionary to China. His friends, Tim and Pam Baker, to honor his memory founded the Philip Hayden Foundation in 1995 for orphans in China, including Shepherd's Field Children's Village with six children's homes caring for more than 4,000 orphans. Gordon Robertson, CBN *WorldReach Founder* calls it a "wonderful work with the orphans of China."

Perry Hayden's unwavering faith produced overtime an immeasurable harvest of souls, not only through Dynamic Kernels experiment, but also by his descendants' ministries, and the Philip Hayden Foundation. Additionally, it is impossible to measure the spiritual impact of the 276 farmers and their church tithe or the impact the Bible-centered harvests attended by some of the greatest industrialist of the time. Moreover, the Purina cereal relief effort kept an unknown number of people alive in Europe. Although Perry Hayden first set out to simply, "demonstrate to the world a fact which appears to have forgotten and ignored," by taking God at his word, God gave him back a heritage of faith beyond anything he could have ever dreamed. The eternal benefits of Perry Hayden's biblical project will only be revealed in heaven.

"Behold, I will make you a new, sharp, threshing instrument which has teeth; you shall thresh the mountains and beat them small, and shall make the hills like chaff. You shall winnow them, and the wind shall carry them away, and the tempest or whirlwind shall scatter them. And you shall rejoice in the Lord, you shall glory in the Holy One of Israel."

Isaiah 41: 15 & 16

Henry Krause

Henry Krause was born the first in a Midwest farm family. Since his father was an invalid he "shouldered the responsibility on the farm and studied the Bible diligently." In 1927, at twenty years old, he moved to Hutchinson, Kansas. Possessing a strong faith in God and intelligence, Henry Krause was inspired to design a piece of farm equipment to counteract the soil erosion problem on the western Kansas windswept plains. In 1916, he saw this was a real need for farmers, and began creating the first One-Way Plow in agricultural history. Over time, word spread and farmers requested his plows that left more soil on the surface in their fields. Known as "God's Plowman," Henry Krause became a respected agricultural leader and innovator.

During World War II the Krause Plow increased the nation's wheat production and food produced from wheat that fed the troops. "By 1947, the Krause one-ways tilled over 19 million acres and helped produce 380 million bushels of wheat!" California dairy farmer and businessman Demos Shakarian once said, "The Krause Plow was used all over the world, and Henry Krause was one of the country's largest manufacturers of farm equipment, and a businessman who devoted half his time and all of his heart to God's service." In 2016, Krause Company will mark its 100th anniversary as the inventor and manufacturer of the first conservation tool in agricultural history.

~Henry Krause, (1907-1972)
Krause Company, Hutchinson, Kansas

Excerpts from The Hutchinson News, Author Amy Bickel

F. Dewey Lockman

In 1927, La Habra, California citrus farmer F. Dewey Lockman surrendered his life to God at a tent meeting sponsored by First Baptist Church of Garden Grove, California. By 1931, the scriptural admonition of Malachi 3:10 captured his heart. He made a lifelong commitment to become a tither. Later, he called this event his "second conversion in the matter of stewardship." Farming citrus and overseeing range acres in productive agricultural community of Orange County region, F. Dewey Lockman and his wife Minna became faithful tithers even when times were financially difficult. On December 3, 1942, with their passionate desire to share the Gospel, they formed the Lockman Foundation with the inspired goal "to reach people around the world." The Lockmans began to sell off their farm in the 1940's as a way to provide sponsorship funds for their Christian literature programs reaching everyone from school children (who called them Grandpa and Grandma) to the military. In 1945, the Lockmans formed their foundation's press and continued reaching souls around the world. During the 1960's, the Lockman Foundation published the Amplified Bible and New American Standard Bible, as well as many other Christian publications.

~F. Dewey Lockman, (1889-1974)
F. Dewey Lockman Foundation
La Habra, California

"Bring the whole tithe into the storehouse, so that there may be food in My house, and test Me now in this," says the Lord of hosts, "if I will not open for you the windows of heaven and pour out for you a blessing until there is no more need."

~Malachi 3:10

Lowell Lundstrom

"Just how do I go about having faith in God? Jesus helps us answer this question: First, He tells you to speak to the mountain in your life. "SAY UNTO THIS MOUNTAIN, Be thou removed and be thou cast into the sea; and shall not doubt in his heart, but shall believe that those things which he SAITH shall come to pass; he shall have whatsoever he SAITH" (Mark 11:23). Whatever your need is –speak to it! …Second, Release your desires! Desire has power to fuel your faith. Jesus said, "Whatsoever YOU DESIRE when you pray, believe that you receive them, and ye shall have them" (Mark 11: 24.) …Don't give up…Third, Forgive. If you have hard feelings against anyone or know of anyone who has bitterness against you, forgive him or ask his forgiveness. If you want a miracle, you need God to work on your behalf-and the only way He'll do this is if you forgive others and keep the channel of forgiveness open. …Have faith in God."

~Lowell Lundstrom, Founder & Senior Pastor
Celebration Church, Minneapolis, Minnesota

Excerpted from Faith for Today

"God of our fathers and our God, give us the faith to believe in the ultimate triumph of righteousness…We pray for the bifocals of faith that see the despair and the need of the hour but also see further on, the patience of God working out His plan in the world He has made… Through Jesus Christ our Lord. Amen"

~Peter Marshall, Chaplian of the US Senate
(1902-1949)

The Heart of Peter Marshall's Faith

Joyce Meyer

"People can wish, wish and wish, but you don't need wish bone, you need back bone. I'm not standing here today because I wished for this ministry, but because I worked hard, fought devils, fought the good fight of faith…fought the fight of faith when my friends rejected me. I fought the fight of faith and made it through getting thrown out of my church, and made it through my family rejecting me. I slept in McDonald's parking lots when I was out on the road for speaking engagements…I refused to quit and I refused to give up."

~Joyce Meyer

Joyce Meyer Ministries
Fenton, Missouri

Enjoy Everyday Life Television Broadcast

Keith Moore

"I'm from a little country town. My wife Phyllis and I had no bank account. We lived in a rickety old trailer out in the country when I began to grow in the Lord. I'd be out in the woods praying, and had no clue I was called to the ministry. Often, I remember being out in the woods in the middle of the night calling out to God in a holy hunger…and that caused me to look up and realize that God had a place for me."

~Keith Moore, Senior Pastor & Founder

Faith Life Church
Keith Moore Ministries
Branson, Missouri

Dr. J. E. Murdock

"Being a farm boy, I did a lot of fence building. I learned by walking along beside my dad why some posts were different than others and why some posts had greater responsibility and were even more important than others. A lot of our posts in the fence are called line posts. And all they had to do is just to hold up the wire. They don't have any other pressures on them but to hold up the wire. But the corner posts are larger and, as my father would say, they are posts that are put there that will last a long time. Where the little line post can be small and can be easily replaced, these corner posts are special…So, in a message we called that a "Corner Post." The Lord has done things, and He still wants to do things in your life that you just can't forget. Can you think of one now?…We call that a "Corner Post."

~J.E. Murdock, Founder, "The Corner Post", Highway and Hedge Ministry

Pastor, Evangelist and Author of, "Truth Out In The Open"
Compiled by Deborah Murdock Johnson

Dr. Mike Murdock

"My Greatest Mind Battle Was Over The Word Of God. When I entered my late teens, intense warfare emerged in my mind. It seemed that contradictions in the Scriptures existed. Then, I began to doubt the validity of the Bible when I saw hypocrisy in believers, inconsistencies in ministers and my own difficulty to live "the Bible standard." Two years of erratic and emotional turmoil occurred. I loved the presence of God and received from the Word of God. However, it seemed that the logic of my mind and the faith in my heart were in constant opposition. One day, in honest desperation, I asked the Lord to provide confidence and inner peace that the bible was truly His infallible Word, not merely the compilation of human thoughts and ideas. The first three powerful truths that emerged changed me forever. Here are facts you should know about the Word of God: No Human Would Have Written A Standard As High As The Scriptures Teach, 2) The Changes That Occur In Those Who Embrace The Word Of God Are Supernatural, 3) The Very Presence Of A Bible Often Produces An Aura And Change In The Atmosphere."

~Dr. Mike Murdock, "The Mike Murdock Collector's Edition,"
The Wisdom Commentary 1;
The Wisdom Center, Denton, Texas

"Fear not earth! Be glad and celebrate! God has done great things. Fear not, wild animals! The fields and meadows are greening up. The trees are bearing fruit again: a bumper crop of fig trees and vines! Children of Zion, celebrate! Be glad in your God. He's giving you a teacher to train you to live right-Teaching the rain out of heaven, showers of words to refresh and nourish your soul, just as he used to do.
And plenty of food for your body-silos full of grain…"

~Joel 1

John Osteen

John Osteen was reared in a poor family, and worked in the cotton fields as a young man. "His parents were cotton farmers and they lost everything they had in the great depression," says his son, Joel Osteen, Pastor of Lakewood. When he was 17, John Osteen surrendered his life to God, and was given a dream in his heart to preach the Gospel. "In the natural he had no future, no hope…(but) God is not limited by your present-day circumstances…He's only limited by our lack of faith," says Pastor Joel Osteen. Although others tried to dissuade him from his dream, John Osteen kept walking by faith. He wasn't satisfied where he was even though he was told picking cotton was all he knew how to do. "He broke out of the old barriers of the past and focused on his dream…and stepped out in faith," says Pastor Osteen. In 1959, John Osteen, and his wife Dodie began Lakewood Church in the East Houston Feed & Hardware. From humble beginnings in an old feed store his ministry and influence for the Gospel grew, and his legacy, and family, now Lakewood Church is considered the largest church in America, reaching millions of people of all nationalities and cultures around the world.

~John Osteen, Founder, Lakewood Church, (1928-1999)
Joel & Victoria Osteen, Pastors, Lakewood Church
Compaq Center, Houston, Texas

Excerpted from *Twice the Life: Looking Beyond Where You Are*

Dottie Rambo

Legendary songwriter and artist, Dottie Rambo was one of the most inspirational Christian songwriters of all time. Of Cherokee heritage she recounted her story of growing up in rural Tennessee, and her fondness for her grandfather Burton visits to their home. Young Dottie would read the Bible to her family while sitting around the kitchen table. Sometimes her grandfather stopped her to share his thoughts, and his heart about a particular scripture. Sometimes if she happened to omit a verse or misread a verse her grandfather would say, "Back up a little, Dottie; I think you missed it!" Under the umbrella of her grandfather's love and his trust in God, Dottie was inspired and grew in her faith. Her grandfather started his evening prayers the same by saying: "Good evening, Lord. Ain't this been a fine day?" Dottie always fell asleep "engulfed in Grandpa's praying."

~Dottie Rambo, Songwriter & Music Legend (1925-2008)
by author Bob Terrell

Christian Farmer's Outreach
Facing the Future with Faithfulness

We go out where they are…it's not our message but the Lord's message.

~ Wilson Lippy, President
Christian Farmer's Outreach

In the Northern Maryland fertile farm country, the Christian Farmers Outreach (CFO) is headquartered, established as a non-denominational ministry in 1987. Both farmers and non-farmers are involved as volunteers and ministry partners who today number more than 2,800 from all walks of life and all ages. At the heart of the CFO ministry is the goal of reaching people for Christ. Attending more than thirty-five agriculture and non-farm events throughout the nation, going wherever the Lord directs, including the North Carolina State Fair, Virginia Farm Show, and the world's largest, California's World Ag-Expo. "The first year we had 1,600 people invite Jesus into their lives," says Wilson Lippy, CFO President and retired generational farmer. Preserving his family's northern Maryland farmhouse built in 1775, Wilson Lippy came to Christ at age 42, formerly farming 10,000 acres. He notes, "My Dad was a farmer and my brothers continue farming." Maryland Dairy farmer John Mike Myers is CFO's Vice-President, John Vogel, Secretary of CFO and farm magazine editor, and CFO Chaplain is Vernon Bolte, who was in agricultural marketing more than 28 years.

"Seven out of ten people never attend a church." This startling CFO statistic is the reason they diligently work spreading the Gospel worldwide. Setting up CFO booths, trained volunteers share one-on-one with a special witnessing tool called the *Bead Ministry*. CFO volunteers bead these bracelets throughout the year. The Bead Ministry was designed to be like the Wordless Book begun more than a hundred years ago and used by Charles Spurgeon, D. L. Moody, Fanny Crosby, Amy Carmichael, and Child Evangelism.

The bracelets are beaded with colors illustrating the plan of salvation: Gold, representing heaven; Black, representing sin; Red, representing the blood of Christ; White, representing all may be saved now; and Green, representing growth in Christ, says CFO. After the presentation by the trained volunteers there is a prayer of salvation. Today, CFO bracelets are on wrists all over the world. "It all works through prayer, through the Holy Spirit," says Wilson Lippy. In 2007, attending the Virginia Farm Show, Jim Bunch was drawn into the CFO booth and listened to the Gospel presentation. He gave his life to the Lord. "Life gets so busy and you just don't make the time you should for Christ," he said. He is now involved in his church's men's ministry.

CFO's biggest challenge is reaching teens for Christ. In his article, "So Many Hear, Then Return With Friends Or Relatives," CFO Ministry Coordinator Arley D. Johnsrud wrote, "They don't want lies, but have great difficulty discerning what's true by looking at today's society…God is surely talking to them. That's why this ministry is so important." One of their most surprising and inspiring teen events took place in February of 2009, when 1500 high school students in North Carolina and California, committed their lives to Jesus Christ. In New Jersey sub-zero weather did not keep high school students from lining up to hear the Gospel. "There are more blessings than challenges," Wilson Lippy says, who has personally trained hundreds of volunteers, 119 for the California World Expo Fair alone. With a passion for its unique mission, CFO helped missionaries and individuals take the Gospel to more than 35 countries in 2008. Equally significant, in 2008, "More than 20,000 people prayed and opened their hearts to Jesus," says Wilson Lippy.

Another part of CFO's ministry is the Deer Meat Ministry program, providing thousands of meals throughout the winter months. In addition, yearly events focusing on the needs of the CFO ministry provide an opportunity for fellowship with farm and non-farm members. Noted guests throughout the years have been former Vice-President Dan Quayle, Joni Eareckson Tada, Dr Ben Carson, Bill Hostentter, who oversees one of the Mid-Atlantic largest grain businesses, and President of Herr Foods, Ed Herr, who spoke on "Committed to Christ" at their 2009 meeting.

Marking 22 years since the founding, CFO has witnessed thousands of individuals praying to receive Jesus Christ as their personal Savior, and continually the organization receives letters from those who came to Christ at their CFO booths. Wilson Lippy sums up their mission by saying, "These people are not flocking to churches--ideally they would. So, we must answer that need by taking the CFO ministry to the people where they are."

"It is written for our sakes because the plowman ought to plow in hope, and the thresher ought to thresh in expectation of a harvest."

~1 Corinthians 9: 10

San Miguel Mission
America's Oldest Church-1625
Santa Fe, New Mexico

Part Five

...through faith they were kept by the power of God

1 Peter 1: 5

Faith on the Farm:
Sowing & Reaping A Harvest of Blessings

Featuring Dr. Oral Roberts

"A harvest cannot happen without first planting the seeds."

~Dr. Oral Roberts

Oklahoma was once a frontier where ancient bands of Indians roamed freely, and buffalo herds grazed the wide, open grasslands. Cherokee, Chickasaw, Choctaw, Creeks and Seminole Indians inhabited the Five Civilized Nation Indian Territory and by 1889 the Five Civilized Nation Indian Territory was opened up to homesteaders. Not long after a family lead by patriarch Amos Pleasant Roberts arrived by wagon. Amos Pleasant Roberts was a fervent Methodist and began farming in Pontotoc County near Bebee. He sowed seeds of faith in God into the lives of struggling pioneer farmers, ranchers, and cowboys after being elected the first superintendent of a Methodist Sunday School in nearby Center. In the years before statehood, he was often called on to preach, and his strong sense of justice was felt in his community. According to author and historian, David Edwin Harrell, Amos Pleasant Roberts was so respected that President Theodore Roosevelt chose him for federal judgeship in the Indian Territory.

By 1901, Amos Pleasant Roberts's son, Ellis Roberts, married Claudius Irwin Roberts, part Cherokee Indian, whose family marched on the Trail of Tears from their ancestral lands in Georgia. Biographer David Edwin Harrell says in *Oral Roberts: An American Life*, one day Claudius rushed through fields to pray for a neighbor's afflicted child before her son, Oral Roberts, was born. Coming to a barbed wire fence, she crawled through the dangerous wire without a scratch and immediately felt the Lord was "hovering near." She vowed to the Lord, if the neighbor child was healed, "I'll give You my son when he's born," recalls Oral Roberts in *Expect A Miracle: My Life and Ministry*. God answered her prayer, and the child was healed.

Granville Oral Roberts was born in a log house on his grandfather's farm on January 24, 1918. Ellis and Claudius Roberts' fifth and last child was dedicated to the Lord as promised. Oral Roberts recalls his mother speaking to him and saying, someday he would preach the Gospel and have a special ministry for suffering people. "A born stutter," he endured agonizing humiliations growing up, but the prayers of faith of his parents finally were answered as a teenager when he was miraculously healed.

Historian David Edwin Harrell says Pentecostal evangelists shook Pontotoc County in 1914. Years later, Claudius Roberts recalled "huge crowds of people" gathered under brush arbors in the lantern light and lifted their voices in praise to God over the Oklahoma rangelands. In 1916, two years before Oral was born, Ellis Roberts sold his farm to pay his debts and "purchased a Bible, a buggy, and a portable organ." He conducted revivals and carried his portable organ miles from his family. Preaching by coal oil light, his brush arbors services drew farmers and ranchers from all over the territory. Ellis Roberts made brush arbors from tree saplings, and trimmed them into poles, sinking the poles in the ground. After a lattice of tree branches was laid over the top of these makeshift shelters, they were complete. "He would then organize a church, secure a young pastor and move to the next place. It was simple but effective," helping them to build a forty by sixty foot building. Without the benefit of modern conveniences or transportation, he established church plants throughout southeastern Oklahoma.

Seasoned farmers, Dr. Oral Roberts recalled working on the farm and in the cotton fields where "the whole family pitched in to pick it." Ellis Roberts loaded the wagon and put his two sons, Oral and Vaden on top, taking his load of cotton to the Ada cotton gin. When he was a young boy, a terrible hailstorm came through their farm and destroyed their crops. "Our loss was devastating," says Dr. Roberts. Nothing was left of their hard-earned seed money, time, and effort including no vegetables for the family's table. "Ellis, you've got to replant," said Claudius Roberts. "It's too late to put more seed in the ground," Ellis said. Claudius reminded him of a twenty-dollar bill she kept in his pocket for safekeeping. After pooling that and the little money in their possession, even Oral and Vaden "farming out" money, they hitched a wagon for Jeter's Feed Store to buy more seed. Although Mr. Jeter said it was too late to plant, Ellis Roberts was unmoved. He put his small amount of money down on the counter. "Instead of telling Papa it was not enough, he told us to drive our wagon around to the back," and take all the seed they needed. Elias Roberts was the only farmer in the area who replanted, and he was the only one who received a crop. Harvesting a bumper crop, he later received double price for their bales of cotton. "We filled our barn with corn, harvested our wheat, and there was great rejoicing in the Robert's house."

Dr. Oral Roberts' ambition as a young man was to become a lawyer and Governor of Oklahoma. But a life-threatening illness of tuberculosis struck him down as a young man a terrible disease that "took my mother's Cherokee people." However, in 1935, through much prayer he was miraculously healed when he was seventeen years old. After surrendering his life to the God, he was several years in the preaching ministry when a local farmer in his congregation changed his ministry.

"I was making $55 dollars a week as a pastor, and I went to prayer meeting under a heavy load," he says, crushed by a financial burden and without a home for his family. Suddenly, at the altar he felt impressed by the Lord to take out his checkbook from his pocket and put it on the altar. "Trust me," he sensed God was saying.

Later, at four o'clock in the morning, a farmer and church member gave him his tithe and said, "This is not just money—it's my seed!" Oral Roberts says, "I'd never heard that before, and I'd never seen a $100 bill in my life! I came out owing nothing, zero." From that point on, Oral Roberts's message on seed-faith giving "came out of the bottom of my heart." Sowing and reaping, faith on the farm molded the life of Oral Roberts. Within a few years his ministry was serving all denominations, and for 40 years he took big tent meetings all over America and throughout the earth.

In 1963, Oral Roberts University was founded by Dr. Oral Roberts and is recognized as one of the greatest evangelical universities in the world. Laying hands on over a half-million people in his lifetime, he says, "I prayed for their healing, laying hands on them regardless of the disease, contagious or not "all manner of sickness…and disease." Dr. Billy Graham, his lifelong friend, to heads of state from all over the globe, the ministry of Dr. Oral Roberts has eternally affected generations around the globe. "Almost every Full Gospel, Word of Faith, and others, have implemented Dr. Roberts's concept of seed-faith giving," says Dr. Paul Crouch. Today after 73 years in the ministry and 61 years in the healing ministry Oral Roberts continues to preach his healing and seed-faith giving message. "A harvest cannot happen without first planting the seeds." From his ancestor's planting, plowing and preaching on the Pontotoc County soil, and his parent's farming, preaching and church building, the life and faith of Dr. Oral Roberts was birthed in the sacred spaces of Oklahoma farmland. Dr. Paul Crouch, Trinity Broadcasting Network President and friend rightly says, "Only when we get to heaven will we know how many have been touched by Oral Roberts Ministry."

"You visit the earth and saturate it with water; You greatly enrich it; the river of God is full of water: You provide them with grain when You have so prepared the earth. You water the field's furrows abundantly, You settle the ridges of it: You make the soil soft with showers, blessing the sprouting of its vegetation. You crown the year with Your bounty and goodness... the luxuriant pastures in the uncultivated country drip with moisture, and the hills gird themselves with joy. The meadows are clothed with, the valleys also are covered with grain, they shout for joy and sing together."

~Psalm 65: 9-13

Pat Robertson

Dr. Pat Robertson was born in 1930, and accepted Christ as his personal savior in 1956. With a mission to "Bring God glory," Dr. Robertson's dedication and commitment to the Gospel and religious broadcasting is unequaled. In 1960, he founded the Christian Broadcasting Network with only seventy dollars, and by faith, established the first Christian television network in the United States. In 1966, he developed and broadcast *The 700 Club*, revolutionizing Christian media and reaching multiple millions around the globe for decades and continues today. A Yale University alumni and lawyer, Dr. Robertson grew up on a former apple orchard in Lexington and his roots are in colonial Virginia. He says: "When I was thirteen years old I went to work on a farm in the state of Virginia. But through that time of physical labor, I grew up and learned to appreciate the land and the labor involved. I was never the same after working in the fields, and have valued that experience all my life."

~Dr. Pat Robertson, Founder and President
Christian Broadcasting Network

The 700 Club & Operation Blessing
Virginia Beach, Virginia

"Observe and consider the ravens; for they neither sow nor reap, they have neither storehouse or barn; and yet God feeds them. Of how much more worth are you than the birds!"

~Jesus Christ, Luke 12: 24

Cheryl Salem

"Mother and Daddy were born and raised in Choctaw County, children of Mississippi farmers… on May 4, 1968, I was crippled as a result of a tragic car accident…I didn't want God to think I didn't have any faith, so lately instead of praying for healing, I'd been thanking Him for my progress…Music was something I was good at, and now-since it was the only thing I could do-I poured my heart into it…"

"I was home setting the table for supper, it occurred to me that I had experienced three things that day that I would not soon forget; first was the thrill of competing, second was the indescribable joy of winning; third, and most important, was the certain knowledge that with God's help even the most hopeless situation can be transformed into a victory! How many times in the past I had heard Daddy quote from the Bible, "All things are possible to him who believes (Mark 9: 13)"

~Cheryl Salem, Miss America 1980
Harry & Cheryl Salem
Salem Family Ministries
Tulsa, Oklahoma

Excerpted, A Bright Shining Place: The Story of a Miracle-Cheryl

Farm Rescue:
Good Samaritan's of the Heartland

"I believe the Scripture not only requires faith, it requires work."

~Bill Gross, Farm Rescue, founder

Jesus spoke of the value of a person's faith, and if their faith was as small as a grain of mustard seed, it could move mountains. Since 2005, the unique vision of one man, Bill Gross, and *Farm Rescue,* has removed mountain-like burdens from the lives of dozens of farmers. Farm Rescue volunteer's assist farmers in crisis within a four state region, North Dakota, South Dakota, western Minnesota and eastern Montana. Growing up on a Cleveland, North Dakota farm, Bill Gross developed the idea for Farm Rescue while flying high above farm fields of his youth as a United Parcel Service 747 pilot. After retirement he'd become, "A Good Samaritan that buys a tractor and goes around and helps farm families plant their crops."

Bill Gross' heart never "left the family farm," and he understood the problems farmers faced, especially the declining numbers of farmers and neighbors to help in their time of need. When a friend challenged him why wait until retirement to help farmers in crisis, he began to put his dream into action. Years ago, farmers helped each other clear their land, build barns and harvest crops. Like a band of brothers, they came to the aide of their farm neighbors in emergencies. Today, with the aging farming population and fewer farming neighbors, Farm Rescue's important mission is a welcome lifesaver for many farmers. Over the past four years, Farm Rescue helped farmers in crisis from car accidents, heart surgeries, cancer recovery, tornados victims and other traumas.

In 2009 alone they assisted 89 farm families in crisis to complete their spring planting season. Farm Rescue's donations come from churches and businesses, including Wal-Mart and RDO Equipment of North Dakota, a family-owned company founded by Ron D. Offett, who grew up on a Red River potato farm. All over our nation's heartland, Farm Rescue's flag is affixed next to the American flag on John Deere tractors rising above the farm fields bringing hope. At this point in time, Bill Gross and volunteers are harvesting fields in North Dakota. Modern-day Good Samaritans, they are sowing seeds of human kindness and love into the lives of struggling farmers and "giving back" to those who give so much to feeding the world.

Dr. Jerry Savelle

"I was born in Mississippi where racism was probably greater there than any state in America. The road where I loved were predominately black people, and I played with black children…The first church I went to was a black church. My Grandmother and Grandfather had a farm so they raised cattle, and hogs, and vegetables. They fed everybody! We had black people at our home all the time. My Grandmother and Grandfather made sure that there was not a black family on our road that ever went without food…One night I heard a scream when I was just a little boy about five years old.

    They were burning a cross in one of our neighbor's front yards, terrorizing that family. With their Klu Klux Klan hoods on, my Dad went back in the house…The man in front said to my Dad, "Jerry, you didn't see anything!" And my Dad recognized his voice- he was the Sheriff in our county and said, "You people come out here bothering these people again, I'll blow you're head off! …They left and they never came back."

~Dr. Jerry Savelle, Jerry Savelle Ministries International
Crowley, Texas

Author of *The Power of Commitment*

R. W. Schambach

R. W. Schambach is regarded as one of the greatest Gospel preachers of our times. When he was just a young man he worked on his uncle's farm, and recalls being awakened before dawn for farm chores. During World War II he enlisted in the United States Navy, and took part in the invasion of Iwo Jima and Okinawa. As a young sailor, R. W. Schambach gave his life to the Lord while aboard ship and preached to shipmates, his pulpit being, "a five-inch gun mount on the destroyer that I served aboard," he says. After his naval service he attended Bible College, and held massive tent revivals all around the nation, and built a church in Brooklyn, New York. One of the most respected and beloved Healing Evangelists of our time, he says, "Faith is the essential prerequisite for everything we receive from the Lord." R. W. Schambach's messages have inspired and changed millions for more than six decades.

~R.W. Schambach,
R. W. Schambach Ministry, Tyler, Texas

Author of Power of Faith for Today's Christian Legacy Series
The Blessing Scriptures, CD Series

San Juan Capistrano
"The Home of the Swallows"
San Juan, California

Dr. Robert Schuler

Robert Schuler grew up on an Iowa farm near Des Moines in the 1930's, and on the Schuler farm there stood a red barn forty feet high. Dr. Schuler says in his book *My Journey, From An Iowa Farm To A Cathedral Of Dreams,* "The barn was cool and dark bearable in summer for both animals and farmers." Inside the barn, hay was stacked high in the loft and the family's riding horse, Mabel, was stabled. However, Dr. Schuler says he did not want to be a farmer. At an early age, he was called of God for ministry. "I didn't choose to be an out-of-place farmer's son. I just was." Dr. Schuler talked to his father inside the barn about his calling for full-time ministry. He dreamed of the day he could attend college, and the day finally arrived when he left his Iowa farm and attended Hope College in Holland, Michigan. Later, the godly young Iowa farm boy was ordained for the ministry. With a solid faith in God and his vision, he began his ministry in California, resulting in the building of the Crystal Cathedral and his broadcast, "Hour of Power."

~ Dr. Robert H. Schuler, Senior Pastor,
Crystal Cathedral in Garden Grove, California

*Author of "Robert H. Schuler: My Journey, From
An Iowa Farm To A Cathedral Of Dreams*

*It is the hardworking farmer who labors to produce
who must be the first partakers of the fruits*

~II Timothy 2:6

Demos Shakarian

"My favorite chore on the farm was to weed the corn because then I could go far in the fields and talk to the Lord out loud. The summers when I was twelve and thirteen the long dim aisles seemed to me like a giant cathedral," dairy farmer, Demos Shakarian said in his book, "The Happiest People on Earth." Demos' father had a dream of owning the largest dairy farm in California, and after many years of hard work and prayer his goal was realized when in the late 1940's as Demos says, "We were the largest private dairy in the world." The Shakarian's relied on the Lord, and called their dairy farm "Reliance." al project the building of Reliance Number Three, the third of our dairies…We bought a forty acre tract and I began the constructions of corrals, silos, and a modern barn and creamery where milk flowed from cow to bottle without ever being touched by hand. And I was forever bringing my Pentecostal beliefs into the cow barns."

~Demos Shakarian, California Dairy Farmer
Founder Full Gospel Businessmen International

"The Happiest People on Earth" by John & Elizabeth Sherrill

Charles F. Stanley

Near Dry Fork, Virginia, Reverend George Stanley, pastor of Emmanuel Pentecostal Holiness Church owned a farm and was a spiritual influence to his grandson, Dr. Charles F. Stanley. At 14, Dr. Stanley received his calling from God, and his grandfather's Virginia farm was a place of spiritual guidance. Today, senior pastor of First Baptist Church of Atlanta, Dr. Stanley reaches millions through his books, tapes and broadcast of "InTouch Ministries." Dr. Stanley's message: *Breaking the Faith Barrier* says, "One of the biggest building blocks to the barrier of faith is fear…" and goes on to encourages us to say to God, "Here I am Lord, I'm willing to trust you, now matter what…"

~Charles F. Stanley, Senior Pastor, First Baptist Church of Atlanta,
Atlanta, Georgia

Author of Breaking The Faith Barrier

Billy Sunday

William A. Sunday was born on November 19, 1862. Not long after, his father, a private in Iowa's Twenty-Third Infantry, Co. E., died during the Civil War. He was raised in an orphan's home. Later became a major league baseball player. After being converted at Chicago's Pacific Garden Rescue Mission, he was called to ministry and became the greatest evangelist of his day. But his heart and messages were close to the earth, he once said: "I was born and bred in old Iowa…I am a hayseed of the hayseeds…out in the log cabin on the frontier of Iowa, I knelt by mother's side. I went back to the old farm some years ago. The scenes had changed about the place…Fingers that used to turn the pages of the Bible were obliterated and the old trees beneath which we boys used to play and swing had been felled by the woodman's axe…I tell you with shame I stretched the elastic bands of my mother's love. I went far into the dark and the wrong until I ceased to hear her prayers or her pleadings…little by little I was drawn away from the yawning abyss…I groped my way out of darkness into the arms of Jesus Christ and I fell on my knees and cried, "God be merciful to me a sinner!"

~Billy Sunday (1863-1935)
Former Baseball Player, Evangelist & Author

Billy Sunday: The Man and His Message by William T. Ellis

Village Missions of North America:

Keeping Country Church's Doors Open

"North America is rapidly becoming the most ignored mission field in the world"

~Brian Wexler, President,
Village Missions of North America

America's small towns and rural communities are the heartbeat of our nation where faith in God and love of family and country are dearly held values for decades. "A vital force of American life," country churches are a large part of our rural vistas as much as white farmhouses, big red barns and rangelands of cattle. Beacons of light, they offer hope and peace to troubled hearts and minds for generations from baby dedications and baptisms to weddings and anniversaries, and the final goodbyes in the church cemetery. The cycle of life for farm and ranch families was a shared community experience, often recorded in historic accounts. Yet, sadly a majority of rural youngsters today will never know these blessings. Along with a growing number of farms and ranches fading from the landscape, along with their demise is the loss of country churches.

    Numbers vary, but experts estimate the rural churches remaining in America, is less than 200,000. Complex issues with no easy answers are being discussed including the changing demographics of our aging generational farmers and ranchers, (average age of farmers today is 56), and tax issues and property values. "Property values and property taxes are rising to the point that some children of family ranchers aren't able to buyout their family's land," says Utah State University professor Dee Von Bailey. The rising numbers of retirees relocating to rural spaces is generating land development concerns, too. In Gallatin County, Montana, 28 ranchers met to talk with County Commissioner Joe Skinner in a one-room schoolhouse about the changing demographics to their region and the proposed "100,000 newcomers" expected to relocate near

Dry Creek within the next thirty years. In these newly developed rural areas it is sometimes easier to demolish the old churches when family connections to the land becomes nonexistent.

"The small rural church is much more than an architectural feature on the countryside…its disappearance would be no trivial matter," said author Eric Sloane more than forty years ago. Today, Sloane's words seem prophetic as officials say denominations are shifting focus from rural churches to urban or suburban churches. Successful Farmer's Magazine reader's survey reported that 74% of farm families attend church, but said: "As farmers become a smaller minority denominations have shifted resources to urban ministries." Isolated pastors serving in rural locales are making it increasingly difficult for congregations to keep the pulpit filled, and, perhaps because for the first time in American history the majority of seminarians do not come from rural areas. According to Village Missions of North America for every church that is opened in North America nearly three churches close each year. Just when they need them most, in many rural places there is not a local pastor to be found. "North America is rapidly becoming the most ignored mission field in the world," says Brian Wechsler, Executive Director Village Missions.

"Do not be anxious for nothing but by prayer and petition, with thanksgiving, present your requests to God" (Philippians 4:6, NIV), is a scriptural promise Village Missions of North America has based their ministry on for more than 60 years. Today this nondenominational ministry has more than 225 communities actively served by Village Missions. Headquartered in Dallas, Oregon, Village Missions is not a church planter but sends trained pastors into existing churches. "They keep country churches alive, " says Brian Wechsler, a native New Yorker. Village Missions of North America is keeping churches a vibrant part of communities in rural mountain regions, farming and ranch states, and western desert vicinities. They have significant results for their labor of taking the Gospel to places where the light is dim or nonexistent. For example, one church with eight members is now seeing 65 attending each Sunday, with more than 25 coming to Christ. In 2007 alone, 700 children and 500 adults were lead to the Lord through Village Missions churches. Executive Director Brian Wechsler says, "Hundreds have become active church leaders, pastors and missionaries, touching the world far beyond their humble home towns."

In the Pacific Northwest, Camano Chapel located on Camano Island, Washington, is a Village Missions church where a community is transformed due to Village Missions. Camano Chapel was founded in the late 1940's after a group of believers sensed their need for a community church. Seattle Post Intelligencer writer Clarence Dirks, football player with the Washington Huskies, wrote about the group building a church and donations came in, enough to complete their church in 1951. In nearby Seattle at the same time, Reverend Billy Graham was finishing a crusade and came to dedicate Camano Chapel. "Thousands gathered for the dedication and many came to know Christ." Reverend Graham's offering was given back to Camano Chapel helping them pay off debts. Over the years the congregation dwindled and in 1967 the leaders decided to close its doors. Someone knew of Village Missions and contacted them. Soon field mission pastors were sent and for years kept the ministry active. In 1997, Pastor Kris and Joyce Kramer arrived at Camano Chapel and since then their ministry has expanded meeting the needs of their community, noted for its lumber and fishing industry and small family farms. "Two years ago we started an evening Bible institute called, Camano Bible Training Center…and we have over 100 students," says Pastor Kris Kramer. Today, Camano Chapel sends six missionary teams to various parts of the world. "We're blessed with great unity in the leadership and congregation," which allows us to train pastors to fill the pulpits of rural churches with a passion to keep country churches doors open.

According to church expert Gary Farley, churches were built six miles from town because the national policy of the 19th century called surveyors to lay out the land in six-mile communities, "It was a six-mile world back then." Near Cadillac, Michigan, the Jennings Community Church is six miles from Lake City and is a Village Missions church. Pastor Larry Shetenhelm, and wife Kathy, are reaching their community for Christ. "My calling here is to be the pastor, and the pastor of the neighboring community," he says. The population of Lake City is around 900, and as an agricultural community is in the midst of transition as there are problems to overcome. "Lake City is home, we grew up here," says Pastor Shetenhelm, who arrived in 2005. "Funds and resources are directed elsewhere and places like rural Jennings and Lake City are overlooked; seminaries and Bible schools don't prepare people for small town ministry anymore." Inspired to walk around his community, as in the Old Testament Book of Nehemiah, Pastor Shetenhelm realized that the problems plaguing urban areas for years were now in rural America. "We're not Mayberry USA anymore," he says. Pastor Shetehelm works with local officials in prison ministries, "I see so much potential in people living in small towns and it truly breaks my heart that their potential is flushed down the toilet with troubles." In spite of these troubling forces, he says, "God is doing a marvelous thing here, and we average 75 a week, ministering to more than 200 community-wide."

A place to worship, a place to pray, a place to relinquish all to God, and a feeling of connectedness, rural churches offer a haven of rest for those living in sparsely populated places. In the small town Formosa in northern Kansas, Pastor Gene Little and his wife, Mary, lead the Formosa Community Church congregation of 45 in an area where there are less than 200. "I went door to door with a deacon in our town and we counted 78 people that live here." On a Sunday morning, the presence of a good percentage of the town is encouraging, says Pastor Gene, who served in four different states and including 20 years in South Dakota.

"God is limitless," wrote *Amazing Grace* author Kathleen Norris. And boundless are the ways God is using Village Missions and drawing people to serve as Village Missionaries. Pastor Dan Clark, and wife, Le Anne of Nisland Independent Community Church, South Dakota, lived in Alaska when God called him to the ministry. "I believe there's a special place in God's heart for rural churches," Pastor Clark says. When his parents died as a young boy he lived on his grandfather's homestead near the town of Wasilla, 60-air miles from Anchorage. "Church for us was the radio," he says. At an early age he developed an appreciation for local churches, "I came to God when I was 15, and God placed a desire in me to serve in ministry." Originally, he planned to be a missionary pilot then God directed him to Village Missions. Pastor Clark writes a newspaper column called "Over Coffee," and set his goal for his church that it was not just "going to survive but was going to thrive." North of Rapid City and near the geo-center of the nation, Nisland's total population is a little more than 200 and Pastor Clark is encouraged with an average attendance of 45, "God has worked in a mighty way in this rural ministry."

One of the difficulties some Village Missionaries face is the naturally independent character of farmers and ranchers who are by upbringing and necessity self-reliant, highly motivated people. This strong sense of responsibility for the land, and ranch, carries over to other areas of life. They naturally feel, "I can do it on my own." Contrary to yielding oneself to Jesus Christ and receiving forgiveness of sin and life everlasting through his shed blood, those traits that makes them self-reliant also keeps them from serving the Lord. Because of this cultural hurdle and the mission field within the heart of our nation, the rural church and pastor who can connect with them in their own world cannot be overstated says rural church experts.

In the heart of the Ozark Mountains near Branson, Pastor Dan Enns shepherds a country church that was ready to close its doors. Providentially, someone knew of Village Missions and not long after Pastor Enns and wife Caroline was sent to the minister at Open Doors Community Church near Cedarcreek. Growing up in a Christian home and reared in the west, Pastor Enns graduated from Frontier School of the Bible in LaGrange, Wyoming. Serving at Open Doors Community Church, he says, "We want them to know somebody loves them, somebody is there if they come in," he says of the transient community. Pastor Enns says the Lord has blessed their church, "My flock is small but I can be personally involved in their lives."

Arizona's Tonto Basin Village Mission church is welcoming and drawing people in the community. "Tonto Basin is a little basin of twenty miles between two mountains, Mt. Ord and Aztec Peak," says Pastor Robert Melotti of Tonto Basin Bible Church, located northeast of Phoenix, Arizona. Pastor Melotti and his wife Stephanie, who grew up on a southern California farm, were lead to Village Missions, "God was calling us to go that route," he says, becoming a Village Missionary in 2001. Seasonal and summertime people are the bulk of the population in this scenic region north of Roosevelt Lake. "We're involved in a local school and minister to those with problems such as alcohol, depression, broken homes and other issues." Tonto Basin Bible Church is meeting spiritual needs, and sharing the ageless message of redemption, forgiveness and restoration in Christ.

Perseverance, patience and prayer are part of a Village Missions life and each mission field requires a heart for the region. "I was intrigued by the BIG things God could do in a small church," says a Village Missions, South Dakota pastor. A small church on a long, paved road with miles of grasslands is Hope Baptist Church on the Montana plains near Ridgeway. Hope Baptist Church is a spiritual oasis to ranchers in this western region, "We have a close-knit community out here," says Pastor Tim Wyrick, who serves as a Village Mission pastor with wife Fern since 2001. Hope Baptist Church is located in the southeast corner of the state near Wyoming, "There are only 80 people in a 400 square mile radius. If we have 40 people in church, that's really good," says Pastor Wyrick, who lives 45 minutes from the nearest town. The distance from the local post office is 27 miles away, and it stands alone on a county road, nothing for miles around. "I like the country and like the fact that Village Missions wants you to spend time with the people." Earning the trust of the ranchers is a big part of his ministry, and he is available to help ranchers at branding time, shipping and shot season, and sheering sheep-whatever they need done. "I'm pretty good with machines so I'm called on to fix trucks. It's neighbors helping neighbors out here." There are about 600 cow and calf operations covering about 14,000 acres going back three generations, and during haying season, Pastor Wyrick works alongside both parishioners and non-parishioners. "I like going to visit them and working with them on their ranches, relationships is the whole basis of my ministry."

Small towns and rural regions need pastors in these days of uncertainty. Today, Village Missions is meeting many of the needs in small towns and rural communities offering hope across North America. Like Hope Church and the others, rural communities are blessed with spiritual guidance and develop their faith in God with a helping hand in time of need, something Village Missionaries say is so meaningful: "That's why I'm here, to develop relationship."

A sower went out to sow, And as he sowed, some seeds feel by the roadside, and the birds came and ate them up, other seeds fell on rocky ground, where they had not much soil: and at once they sprang up, because they had no depth of soil. But when the sun rose, they were scorched, because they had no root, they dried up and withered away, some fell among thorns, and the thorns grew up and choked them out. Other seeds fell on good soil, and yielded grain-some a hundred times as much as was sown, some sixty times as much, and some thirty….As for what was sown on thin (rocky) soil, this is he who hears the Word and at once welcomes and accepts it with joy: …As for what was sown among thorns, this is he who hears the Word, but the cares of the world and the pleasure and delight and glamour and deceitfulness of riches choke and suffocate the Word, and it yields no fruit. As for what was sown on good soil, this is he who hears the Word and grasps and comprehends it; he indeed bears fruit and yields in one case a hundred times as much as was sown, to another sixty times as much, and in another thirty."

~Jesus Christ, Matthew 13:3-23

Chisholm Fork Cowboy Chapel
Beaumont Ranch
Fort Worth, Texas

Part Six

*...I live by faith in the Son of God,
Who loved me and gave Himself up for me*

Galatians 2:20

America's Cowboy Churches

Barns, Chapels & Cowboy Faith

"I've been helping with cowboy churches since 1984 when I found I could serve the Lord with my music."

~Susie Luchsinger, Cowboy Church, RFD-TV

"No fingers are pointed in judgment…no one should buy a new outfit…a place of solace… and a place to rejoice in the Lord," is just part of the code of *Chisholm Fork Cowboy Chapel* that stands on Beaumont Ranch's *Chisholm Fork Old West Town* near Ft. Worth, Texas. Chisholm Fork Cowboy Chapel was built in 1999 and created and owned by Ron and Linda Beaumont to reflect the western chapels of the Old Chisholm Trail of the 1890's. Dusty boots and faded chaps, the cowboys of the late 19th century walked into houses of worship like Chisholm Fork Cowboy Chapel with their hats over their hearts and listened to frontier preachers deliver the Word of the Lord. The Old Chisholm Trail was named for a historic figure named Jesse Chisholm, who owned a trading post on the same route where cattlemen drove cattle to market. "Each town of that time period had the church at one end and the saloon at the other," says Linda Beaumont, who with her family manages the 800-acre ranch that includes a 20,000 square foot Lone Star Barn.

Chaplain John Moore, formerly with the Billy Graham Evangelistic Association, shepherded for many years the small flock that would gather at the Chisholm Fork Cowboy Chapel seating forty. Distinctive, the chapel is affixed with a wooden front door that holds a colorful mosaic window and the roof is topped with a bell tower and a large cross. A place of solace, it is also an unpretentious place where anyone can come and, "rejoice in the Lord and his works." Chisholm Fork Cowboy Chapel is set apart on the landscape, and speaks of former days when the west was wild but hearts were always welcome at the Lord's Table. Now, after a decade of dedication the family-owned chapel is one of the first to honor the faith of the old west cowboy.

Today, the state of Texas ranks number one in cowboy churches in America. Cowboy Churches have been defined as places where money, prestige and material possessions aren't the measure of success. In rural America, one of the fastest growing Christian outreaches is cowboy church. Ellis County, Texas has the largest cowboy church in the nation called *Cowboy Church of Ellis County,* with 2,000 attending. Generally, they are established by men and women who are "one of them," cowboys and cowgirls who have a history of ranching or farming, or with the rodeo circuit. Horse enthusiasts, and those associated with horse ranching, are also growing in numbers and are finding a home in cowboy churches. "Yes, Lord, I'll ride with you," is being repeated across the country within networks of cowboy churches from various denominations. The *Texas Fellowship of Cowboy Churches,* in conjunction with the Baptist General Convention of Texas, is considered the leading outreach with 84 cowboy churches and growing. Headquartered in Waxahachie, Texas, Texas Fellowship of Cowboy Churches Executive Director, Ron Nolen started *Cowboy Church of Ellis County* in 2000, with wife Jane, and served as pastor. Currently, plans are underway for another 200 church plants.

Riding with the Lord for many years is professional rodeo circuit rider Tony Shoulders, founder of *Rodeo Ministries,* Emory, Texas, who answered the call to ministry in 1994. Tony Shoulders thought Jesus was a "sissy" until he was born again at a cowboy church service in Branson, Missouri in 1980. Committing his life to serve the Lord after a near-death injury. Now using his testimony with wife Cindy, at every opportunity and as a member of a championship rodeo family and professional rodeo circuit rider he says, "Praise God for men…who plowed some mighty rough ground in order to plant the seed of cowboy churches."

America's cowboy churches are now in nearly every state and are multiplying experts say because ministry leaders understand the cowboy culture and they empathize with their demanding and vitally important lifestyle. For example, *Cowboy Church of Oregon* meets at The Eugene Livestock Auction Barn near Junction City, where Pastor Tom Crabb's says it is "a rescue mission just a mile from hell reaching out to all who are lost." In Vinita, Oklahoma, *Cowboy Junction,* founded by Wade Markham and his wife Louise, began inside their barn in 1989. Wade Markum was a professional rodeo cowboy and started with a Bible study for cowboys. Now his whole family is part of the ministry at "Cowboy Junction" and puts "people in touch with the God of Hope." In Lee County, Mississippi, a barn is where Pastor Ken Pollack's *Cowboy Church* where singing praises to the Lord and answers to prayer are a large part of the service; like a young girl thanking "Jesus for my recent cancer victory" and singing Martina McBride's "Anyway."

In the 1980's, *Cowboys for Christ,* founder Ted Presley began his ministry that influenced thousands. Farmers and ranchers, and those associated with ranch-life are being ministered to in unique, scriptural ways through cowboy churches. Rugged individualists with long work hours and living great distances from the nearest church are coming to cowboy churches and know they won't feel guilty or excluded if they can't make it to a service because of their demanding professions. "When its planting time or harvest, you can't just up and leave the farm or ranch."

For decades agricultural communities were linked in a bond of community by necessity of clearing the land, raising barns, building churches, holding quilting bees, and baptisms in the nearby creek, helped families grow up together for generations. Today, ranchers and farmers, and those who love the rural lifestyle are brought together through RFD-TV, founded by Patrick Gotsch, and the launching of a new program called *Cowboy Church* has increased public awareness and love for the rural outreach sweeping America. In September 2007, Cowboy Church, RFD-TV team began, lead by producer, Jim Odle, and hosted by Gospel music artist, Susie McEntire Luchsinger, and co-host Russ Weaver, pastor of *Shepherd's Valley Cowboy Church*, Egan, Texas. "This team travels across the country to lead a unique worship experience," says RFD-TV, *The Magazine*.

"Countless blessings and testimonies because of Cowboy Church's message," says Jim Odle. A big dream fulfilled for Jim Odle, an auction owner, who sold his company the "Superior Livestock Video Auction" in order to free time to launch Cowboy Church. Cowboy Church-RFD TV reaches now around the world bringing the redemptive message of Jesus Christ to those who have never entered a church. "We have a great opportunity to turn around the decline of Christianity in rural America," says Pastor Russ Weaver, an Assembly of God missionary. Pastor Weaver and his wife Anna, oversee Shepherd's Valley Cowboy Church. "Our goal is to bring the gospel and encouragement to those in the cowboy culture," says Pastor Russ Weaver of Cowboy Church RFD-TV, who roped calves professionally for many years. "We praise God for those who contribute to the show on a monthly basis. It sure helps us and encourages us to keep on," says Susie McEntire Luchsinger, whose headquarters are in Atoka, Oklahoma.

Cowboys, ranchers and farmers say their reward is just being a part of the landscape and doing what they love. Since the early 1980's, Gospel Artist Susie McEntire Luchsinger has been doing what she loves, and what God has called her to do, and that is ministering at Cowboy Churches. "I've been helping with cowboy churches since 1984 when I found I could serve the Lord with my music," says Susie Luchsinger. Multi-award winning Gospel artist, she released her 15th album "Let Go" in 2008 and is co-writer of "Sticks and Stones" with niece Autumn McEntire Sizemore. Combining her rural cowboy ancestry and upbringing, she shares the Gospel of Jesus Christ and say, "I knew in my heart God had called me to sing, and He promised I'd never lack for a place to sing-- and in 25 years God has kept me busy!" Reared in Atoka County, in southeastern Oklahoma, Susie McEntire Luchsinger is a third generation ranch cowgirl. Her grandfather, John McEntire, was a world champion steer roping holder, and father, Clark McEntire, won titles in the same event in 1957, 1958 and 1961.

"At times being a rancher, it seems like you have everything going against you. Ranchers have to be really creative to survive," says Jim Ballard, Director of Missions East Idaho Region covering Idaho, Montana, Colorado and Wyoming. "Blue jeans and boot culture make up much of the states of Idaho, Montana, Wyoming and Colorado where I serve and these rural people can't really identify with a traditional church. Ranchers branding calves, and potatoes farmers can come straight from ranch work and chores feeling comfortable in a cowboy church." Pastor of Blackfoot, Idaho's *Christ's Country Cowboy Church,* Jim Ballard's 37 years in ministry began with a call to preach at 15, "I worked on farms and ranches all my life," he says. With the awe-inspiring backdrop of the Blackfoot River and Salmon River region, he says, "Some of the most beautiful rivers in the world are here, and much of Idaho reminds me of Colorado when I was a kid." Growing up on a Colorado farm, Jim Ballard later ranched and understands the ups and downs of both ranching and farming. As a full-time missionary, he travels an average of 400 miles on Sundays to preach and oversee church plantings. "I'm meeting and training church planters all the time, and we've had three churches start in the past year." With wife Myrtle, he says, "If you care about folks, you're going to spend time with them and listen to what they have to say."

Lemhi River Cowboy Church, Salmon, Idaho, is one of several cowboy church plants directed by Jim Ballard, "What God has done with Lemhi River Cowboy Church is truly a powerful thing." In 2006, Pastor Mike Palmer established Lemhi River Cowboy Church, and 90 are in the congregation today, with 40 attending Vacation Bible School. "One 70 year old cowboy came to Christ there who never went to church before," says Jim Ballard. Inspired by the fruit of their labor, other churches are planned like one for Native Americans. "Church planting is my heartbeat," Jim Ballard says.

Cowboy faith is on the rise in rural America. "Rescue the perishing" is not just lines of a classic hymn, but the heartfelt call of pastors serving cowboys and ranchers. "Wyoming is a tough area," says Pastor Deanne Graves of *Hilltop Baptist Church* in Green River, Wyoming. One of the biggest challenges facing Pastor Graves is that cowboying is a transitory community. "They come here to work for a while and then move away back to their ranches." Growing up working cattle in northern Colorado, Pastor Graves worked on ranches all his life and is of the land and understands the culture. "I can do this on my own-no need for Christ," is a prevailing attitude that Pastor Graves says keeps many from coming to church. Unmistakable, Pastor Graves knew God had called him to Green River, the point of beginning for John Wesley Powells 1869's expedition. Pastor Graves says, "My wife and I had a ranch and we're happy, but then I sensed God was calling me to step out in faith. So I said, 'Ok, I'll take the call if You provide for our livestock.' That day three people phoned out of the blue and offered their ranches making us free to move. God made it clear I was to accept this ministry call."

"Cowboys, and those associated with the cowboy culture, don't want the Gospel sugarcoated," says Jay Avant, pastor of Milltown Cowboy Church near Davenport, Florida, "They want the plain truth, and they want it to where they can understand it." Florida ranks first in the nation with the most ranches, a surprise to many, with approximately 18,000 cattle ranches.

The *Double Bar C Ranch,* is a Brahman Cattle Ranch near St. Cloud, and owned by James and Leslie Chapman and their family. Committed Christians, the Chapman's serve at *Canoe Creek Christian Church.* "I'm an elder and adult Bible school teacher and my wife Leslie, heads up the nursery and small children area," says James Chapman, whose family and ranch family are active in church. The Double Bar C Ranch's web site features scriptures, and James Chapman says many around the world have responded to them. "Its such a blessing to receive comments from countries," and he adds, "God works in so many different ways that we are not aware of." Like the Double Bar C Ranch, God is working in many ways in Florida reaching the cowboy culture through Cowboy Churches. "Florida has a strong ranch culture, and is going to be one of the great states in cowboy churches," says Pastor Smith founder and Executive Director of *Cowboy Church Network of North America*. "Bringing Jesus to rodeo riders, barrel racers, ranchers and farmers," Cowboy Church Network of North America baptize in horse troughs and diligently working to train and equip churches in their network and young pastors called to cowboy church ministry.

"Brass Tacks for Cowboy Church Planting," is a 4-hour course offered free of charge by Pastor Smith. "Our biggest need is more preachers that can reach the cowboy culture," he says, conducting training sessions for young pastor's who are answering the call to reach the cowboy culture. One young man asked Pastor Smith to pray that his father who was on drugs would come to cowboy church, and later their prayers were answered as his middle-age father came to church and was saved, only to die that week.

"Cowboy churches are lights in North America," says Pastor Smith, whose ministry is based in North Carolina and in association with the North American Missions Board of the Southern Baptist Convention. "Cowboy churches are real churches with bylaws, and we baptize, marry and bury, and minister to the spiritual needs." Pastor Smith began this ministry, "just trying to be good father for my daughter and buying her a horse eleven years ago." But while talking to folks at feed stores and on trail rides, as a traditional pastor inviting them to church they declined. "'I'm just as good as those fancy people in traditional churches,'" they'd say but later when I asked them if they'd come to a Monday night gathering in our barn, they said, 'Yeah, I'll come.'" Since 2004, the Cowboy Church Network of North America has seen hundreds come to Christ and faithfully serving God, and reaching the cowboy culture now with more than 60 churches. "Our goal is to impact the cowboy culture for Jesus Christ in every county, province and territory of North America." Often the question posed to Pastor Smith is: "Can you have church in a barn?" and he answers by reminding them that Jesus was born in a barn and wore carpenters clothing for his trade. "We're wearing jeans and boots, part of our trade," says Pastor Smith, who participates in "shoot-dogging" at events and trains his own horses. Skeptics are invited to visit cowboy churches, "I say to them please come to cowboy church one time, and when they do they realize we're very serious and people are getting saved." Far-reaching and with a vision for cowboy churches to be, "A light in every barn," Cowboy Church Network of North America are meeting in barns, horse arenas, on trail rides and in pastures. Often traditional churches send the message, "You come and look like us," but in cowboy churches they are reaching people as missionaries, as in other cultures, and saying, "We want to come to you'all, in your cowboy culture."

Circle K Arena, Mt. Pleasant, North Carolina, and Central Station Church, Midland, North Carolina are two ministry outreach centers, with a Monday night meeting he calls "The Gathering," for folks to come and fellowship with more than 60 attending. Vacation Bible School videos produced by LifeWay, Nashville, Tennessee, features Pastor Smith's Cowboy Church Network of North America, and his input guided the songwriters for the video, "Saddle Ridge Ranch."

PBR, the world's biggest bull-riding event held in Winston-Salem, North Carolina is another ministry outreach, and "Cowboy Church Roundup" is held each November celebrating all involved in Cowboy Church Network of North America. "We have a big rodeo and it's very exciting with folks come from all over to be baptized in a horse trough, and I give an invitation and ask if there is anyone who wants to accept Jesus Christ and they are baptized also." (Ministry events located on their website, see endnotes). "Our goal is to impact the cowboy culture for Jesus Christ in every county, province and territory of North America," says Pastor Smith.

*"I remembered the old days, went over all you've done
pondered the ways you've worked, Stretched out my hands to you, as
thirsty for you as a desert thirsty for rain…
Teach me how to live to please you because you're my God.
Lead me by your blessed Spirit into cleared and level pastureland."*

Psalm 143
Message

Frank & Lola Yassie
Evangelists to the Navajo Nation for 60 Years

New Mexico

The Dineh, the Navajo's

Desert Camp Meetings, Miracles & Ministries

"We drove to far areas to set up camp meetings"
~ Lola and Frank Yassie, Navajo Evangelist

The Navajo Nation encompasses a four state region, namely Arizona, Utah, Colorado, and New Mexico, and is the largest Native American population in the United States. With nearly 300,000 inhabitants, its landmass of 17.5 million acres holds red sandstone mesas and miles of breathtaking wonders and home of *The Dineh*, the Navajo. Earth, wind and sky witnessed the forced removal and exodus of the Navajo's in the 1860's called, "The Long Walk." Although they were cast down and displaced from their ancestral territories, the Great Spirit raised them up, and their presence blessed the landscape. The *Dineh,* meaning *The People* in Navajo, are an intelligent and gifted culture, recognized for their remarkable creativity and beauty.

Under jade-blue skies, they hand-weave colorful blankets on looms with patterns unique to their individual experiences. "Navajo weavers bring life to their products," says expert Teresa J. Wilkins, author of *Patterns of Exchange*. Acts of worship to their Creator, their designs are a sacred part of their existence and commonly described as "their thinking," says Teresa J. Wilkins. Besides creating turquoise and silver jewelry, pottery and basketry, the Navajo's enjoy gathering around desert campfires safe and warm. Ancestral stories around these campfires often recall the Long Walk and the setting of reservation-born, Frank and Lola Yassie, who were reared near Gallup, New Mexico. Evangelists to their fellow Navajo, the Yassie's were one of the first to evangelize among their people, traveling thousands of miles to desolate places. Born on the reservation in 1914, Lola Yassie was stirred early in life by what she calls her visionary dreams.

"There were no preachers and no churches nearby in those days, and it is evident the Holy Spirit spoke to her heart drawing her to the Lord," says Pastors Martin and Christine Kirk, *New*

Life Pentecostal Church of God, of Thoreau, New Mexico. As a young woman, Lola sensed God was showing her the spiritual needs of her Navajo people and was burdened for their souls. Lola's faith in the Christian God grew in the quietness of her family's horse ranch without discipleship or training of any kind, and with her family of the traditional Navajo religion. But then a tragic accident severely tried Lola's faith. "I was on the ranch when some horses ran away, and I got on my horse and ran after them." In the confusion, Lola was thrown to the ground trying to recover the lost animals. Lola lay motionless with her back broken, and she says her family thought she was dead. "I was in the hospital a long time but I still suffered and they said I had no hope for permanent recovery."

    Not long after, one of her Christian family members learned that a man named Oral Roberts was coming to Albuquerque, New Mexico. Hopeful, her family carefully drove her to the tent meeting. After arriving, they heard thousands praising God, and Lola says, "We could only squeeze way in the back of the tent and I waited for prayer." In great pain, Lola remembers, "I was listening to his message and hurting so bad." Then, as she lay still she began to feel a change in her body and her faith in God's healing power was strengthened. "All of a sudden I was feeling better. No one touched me that day but as I lay there God healed my body." While the preaching continued Lola energetically recalls today, "I was healed from then on! And I never had anything else happen to my body, and I was so happy!" A miracle, Lola began to share her story, and her commitment to God offered hope to other Navajo's.

    Frank Yassie was born in 1912 in northwestern New Mexico, and was led to the Lord by his future wife, Lola. Orphaned as a boy, Frank Yassie lived on the reservation until serving during WW II as one of the 44,000 Navajos from 1940-1945. An Army paratrooper, Frank Yassie was hitching a ride on a mail truck back home to the reservation while on leave, when he met Lola who was sitting in the cab of the truck returning home from school. Later, after marriage Frank Yassie served as a New Mexico Range Patrolman. During those lean, post-war years, the Yassie's lived on the reservation and reared their family living trusting God to provide.

    "Navajos call the years immediately following World War II the 'starving time,'" says Native American author Kathleen P. Chamberlain in, *Under Sacred Ground: A History of Navajo Oil.* Surrounded by the Zuni Mountains, Frank and Lola Yassie carried the Gospel and ministered in desolate places and share the little they possessed. Lola, 95, says, "We'd all pile into our wagon with our sheep and drive to far areas to set up camp meetings." Sheep is a staple in the Navajo diet was their only food supply, and around revival camp meeting fires they'd roast the sheep and eat it with Indian flat bread, says Lola. Within this rocky, mountainous region with sandstone cliffs, the Yassie's saw the suffering of their people, both materially and spiritually. They knew God was the answer to the hurting Dineh around them who came to listen to this consecrated couple and Lola's story of a miraculously healing. No distance was too great to share God's Word. In spiritually barren areas and from great distances the Navajo's came by foot, horses, and trucks eager to hear their message. In 1953 their confidence in their calling was supported when they attended a meeting of evangelist, A. A. Allan, a world-renown preacher of the day. Out of the crowd he summoned Lola and Frank Yassie, and prayed over them. "He said they would greatly influence their generation for the Gospel," says Pastor Kirk. Conflict with the traditional Navajo religion was one of the strongest trials for the Yassie's, Lola says in broken, but clear English, "We encountered many hard things."

After a life of service to God, Frank Yassie passed away in 2009. He was a respected Biblical teacher, and was legally blind unable to travel before his death. Today, Lola continues to be a fervent woman of prayer after the "home-going" in 2009. "She is a great woman of God," says Pastor Kirk. An inspiration to Navajo women, says Christine Kirk says, "Lola still climbs trees to pick peaches!"

Life on the Navajo reservation has not changed much as it was decades ago when Frank and Lola grew up there. Today, experts say most Navajos live below the poverty rate with approximately median income of less than five thousand dollars. This can be seen in the life of a Navajo woman named Mary, age 80, who was born in a log Hogan and lived there all her life. She was dying when the Kirk's were asked to visit her, and after prayer, she was miraculously released from the hospital. Mary never had running water or a bathroom, and collected her own wood for heat. Like Mary, in many areas, the Navajo are living without the bare necessities of life, including water and sanitation. "Many people get their water from water pumping stations, and some still carry water in buckets as their ancestors did one hundred years ago," says Pastor Kirk, who minister to non-church Navajo's as well, like a regional Navajo shepherdess in her nineties who ran a small family ranch. "No grass on the landscape and having to haul water for their animals and the well where they get their water is inside the reservation nine miles from their home," says the New Life Pentecostal Church of God newsletter.

Unemployment and alcoholism, depression, despair, and death are the devastating effects on the reservations. Alcoholism affects 97% of the people, "There is a high rate of deaths among youth on the reservation mainly due to alcohol and few jobs," say the Kirks, who work with professionals at Gallup, New Mexico hospital. A former professional truck driver, Pastor Kirk was traveling on I-40, through the Thoreau, New Mexico area when he says, "God spoke my heart and said, 'You're going to pastor a church there.'" The Kirks left their Michigan home to minister fulltime to the Navajo people in 2005. They instill renewed pride in their Navajo parishioners' culture and talents and build their faith in God's promises, "We tell our church family, 'THINK BIG and don't settle for a little." Christine Kirk conducts classes on canning peaches and other topics of interest to the Navajo women, and organizes Mother's Day and after giving each Navajo woman a long stemmed carnation she says, "They were so thrilled because they'd never had a long stemmed carnation given to them before."

Serving the Grand Canyon, Monument Valley and Lake Powell area of Arizona since 1940 is the Flagstaff Mission to the Navajo, Flagstaff, Arizona. The Flagstaff Mission to the Navajo is an interdenominational ministry with a goal to reach the Navajo, the Dineh, for Christ on the reservation. "We have eight churches on the reservation," say Pastor Mike Calvin of Flagstaff Indian Bible Church, one of the churches of the Flagstaff Mission to the Navajo. The mission has grown with several pastors and support teams ministering to the needs of those on the reservation within the region since founders, Katherine Beard, Imo Wardlow and R. Walter L. Wilson, established the mission in the 1940's.

One of the challenges facing the ministry over the years is again the difference between Christianity and the traditional Navajo religion. Today, many consider suffering for the Gospel is something only experienced in foreign countries, but not in America. However, many Navajo's risk great distress and ostracism from family and friends after surrendering their lives to Christ, no longer practicing the Navajo religion, "They cast us off, they don't associate with us," says Pastor Calvin. Pastor Calvin, the origins of the Navajo as taught are myths contrary to the Bible. "As Christians, we believe the Bible and come from that prospective," says Pastor Calvin.

Navajo Pastor Scott Franklin of the Flagstaff Indian Bible Church is an exception to this dilemma, and was supported by his father when he became a Christian. "In 1953, I dedicated my life on the basis of Romans 12:1 and surrendered my life to the Lord," Pastor Franklin says, who ministered to thousands of Navajo's traveling throughout New Mexico, Utah, Colorado, Arizona, and other states. "You lead, I'll follow," was Pastor Franklins' prayer to God as he stood at the foot of a local mountain near Flagstaff, and left a good job as a clerk at an ammunition plant to preach the Gospel. "I still have my King James Bible from that time," he says. Born on the Navajo reservation in Arizona in 1921, Pastor Franklin's father taught him how to live a righteous life and was a well-respected man on the reservation.

When he was five years old, his father was taught him how to tend sheep, "At six years old I was shepherding my family's large flock." As he matured, he began to sense he needed something more than what the Navajo religion offered. "The more I thought about it, the more I began to see that the real truth must be found in the Holy Bible of the Christians." After Bible school and working with the Arizona Baptist Convention for decades, Pastor Franklin was frequently asked to start new churches among the Navajo, a new concept back then. "We started five churches, and the Navajo's started running their own churches." Camp meetings in the desert are one of the avenues of ministry to the Navajo who are by culture a nomadic people. The act of traveling is to them a sacred thing in life. Speaking at camp meetings over the years, Pastor Franklin's ministry was groundbreaking for the time, inspiring his people to leadership. "I was an example to other Navajo's at camp meetings and preaching at more than 60 churches." Now, at 89, he is a spiritual advisor for The Flagstaff Mission to the Navajo, and says only the Lord knows how many came to God through his ministry.

Flagstaff Missions to the Navajo is one of the largest ministries serving the Navajo's. Others are diligently working to uplift and encourage the Navajo's, and other native bands of First Peoples, offering them the hope and peace and to let them know they are not forgotten. "Some think of missions and cross-cultural ministries only in terms of going to foreign lands. Yet, opportunities abound in our own backyard and many times at less expense, including outreach to the intercity and rural reservations of the Native Americans," says Flagstaff Missions to the Navajo.

Dyer Memorial Chapel

South Park City, Colorado

"Then He showed me the river whose waters give life, sparkling like crystal, flowing out from the throne of God and of the Lamb…on either side of the river was the tree of life with its twelve varieties of fruit, yielding each month its fresh crop; and the leaves of the tree were for the healing and the restoration of the nations."

~Revelations 22:1

Acknowledgements

"When God calls you to a great task, He provides you with the strength to accomplish what he has called you to do," wrote my heroine, Coretta Scott King. God truly strengthened me to complete this book, and I acknowledge Him first and thank Him for all the wonderful people making it possible. My gratitude is extended to those who understood my vision, especially: Dr. Stephen G. Scholle, and Heather L. VanKoughnett of the Billy Graham Evangelistic Association; Jeff Burton, President of ITEC Entertainment, designer of the Billy Graham Library, and Trish Loubier, Director of Marketing for ITEC. Thanks to all ministry staff handling copyright permissions over the years, including, Barbara Johnson, Executive Assistant to Dr. Pat Robertson; Pastor Jerry Falwell, Jr. for special permissions, Abigail Ruth Sattler, Liberty University Archivist, and Liberty University staff for the Worley Chapel photograph; John Copeland for Kenneth Copeland Ministries and Cindy Harnes; "Dr. Paul Crouch's Executive Assistant Margie Tuccillo, TBN, and Ms. Ruth, TBN Hendersonville church, (TN); Pat Harrison for Kenneth Hagin Ministries; Annette Garcia and the Gonzales family, Sun Broadcasting;; Tina Ketterling, John Hagee Ministries; Jim & Vicki Mann, Little Brown Church, for interviews and copyrighted photograph permission; Alfred E. Brumley Music for permissions; Dr. Trevino, for Dr. Mike Murdock and The Wisdom Center; and Peter J. Daniels, Australian farmer and businessman for his email prayer. I'm grateful to Wilson Lippy for reviewing CCF chapter, the ORCF review, and thanks to Pastor Jimmy Snow for his story. I'm also very grateful to Stan Hitchcock for his interview and his inspiring country church video. My sincerest appreciation to all the ministries entrusting their stories to my care, I'm honored.

Gratitude to my precious friends, Pastor Keith & Brenda McAra, Worldwide Harvest Church, for their faithful prayers over the years; Reverend Joan M. Brown, Founder, Heart to Heart Ministry, for her wonderful friendship, faith-filled support and anointed prayers incredibly adding to the vision of this project, without which this book definitely would not be in your hands; Ron and Rebecca Hamming, for their dear friendship especially at a pivotal time; appreciation goes to Pastor Tim & Rhonda Forstoff, Cornerstone Church for their their great ministry and wonderful messages keeping the vision alive for this book; and thanks to the ministry of The Outpouring Cafe, Paw Paw, Michigan, and my sisters in the faith, Taye and Heidi, and their families, for their vision and spirit that has been such a blessing to me (and great food!). Thanks to Linda Parmer, Stonecroft Ministries, International and directors at Stonecroft; Stephanie V. Slavin, Quantum Leap, Inc., my mentor and wise friendship; Dr. Dennis E. Hensley, Professor of English, mentor and editor for going the extra mile and for with chapter reviews, thanks, Doc (any mistakes are my own); Polly Wirtz, PeakView Consulting, Inc. for her valuable input; Barry & Virginia Kerrigan for their expertise and time, incredibly capturing the message of my book; Sue Balcer, for her great work, and Rod Humiecki, Titan Photo Lab, many thanks. I'll always be grateful to fellow press woman, Willah Weddon, for mentorship and the Weddon family.

    Thanks to Linda Beaumont, Beaumont Ranch, for copyrighted photographs permissions and support. I appreciate the following people for their help and encouragement. From the Governor's office and State Representative Dave Robertson on down I'm grateful for all those supporting my rural preservation endeavors, and Ray Bergdolt, Wm. Tiny Zehnder Foundation, The Perry Hayden Family, Betty Poe Henry, Brian Phieffer's Lawn & Garden Tractor Magazine, and PBS Michigan Magazine's Barry Stutesman & Dell Vaughn. Researching this project took many years and thanks to the staff of the centers including: Benson Ford Research Center, Tecumseh Library, Library of Michigan Flower Pentecostal Archives, Patricia Margosian Terrell, Director of Communications for San Juan Capistrano; Sheila Bumgarner, Public Library of Charlotte & Mecklenburg County, Barbara Dey Colorado Historical Society Specialist. My deep gratitude to Bobbie & Frankie Tucker for their gracious hospitality and support; appreciation to Loretta & Paul Brown, Lloyd & Jane Ervin, and Jim & Katrina Ervin, for their warm hospitality and understanding. My love and thanks to Priscilla Shoeman, the first one to share God's love with me on an airplane in 1970, and husband, Pastor Bruce Shoeman. Paula Carpenter, Sue & John Smith, and Dave Clark, my mentors, only God can reward you for all you've done for me. Charles Capps booklet and words: "Stay with what God has anointed you to do, stay with your calling in life," brought great momentum to project, thank you. My thanks and appreciation to my husband John, and sons, Andy and Rob, men of great faith for their belief in this project and their sustaining prayers.

Endnotes

Introduction:
Just As I Am: The Autobiography of Billy Graham, pg. 23 & 24
A Time for Remembering: The Ruth Bell Graham Story, pg. 52
Mecklenburg-Charlotte Library: Mecklenburg by Howard & Ruth White, pg. 258 Sharing Our Heritage, Sharon Township, pg. 41
Charlotte-Mecklenburg Historical Landmarks Commission
Billy Graham Evangelistic Association Official History
Author's Archives: MSMBC

Faith in the Country:
The Rise of Evangelicalism by Mark Noll, pg.77
American Colonies: The Settling of North America by Alan Taylor, pg. 348 & 349
The Reshaping of Everyday Life, 1790-1840 by Jack Larkin, pg. 278
Southern Evangelicals and the Social Order by Anne C. Loveland, pg. 68
Christ in the Camp by Chaplain by J. William Jones, pgs. 171,306, 391
United Methodist Archives & History: United Methodist Historic Shrine
Author Archives Letter from Betty Poe Henry, 2002
Old McKendree History, courtesy Old McKendree Board of Trustees
Stan Hitchcock's "The Country Church Songbook Collection": www. BlueHighwaysTV.com
The Azuza Street Revival by Bishop George D. McKinney, University of Alabama (Lecture, 2001 "Pilgrims of the Sawdust Trail")
International Pentecostal Holiness Church History "1901 to the Present"
The Azuza Street Mission and Revival by Cecil M. Robeck
Flower Pentecostal Center, Springfield, Missouri

Little Brown Church in the Vale
Dr. William Pitts Recollections; Chickasaw County: Famous Chickasawians
Interviews with Pastor Jim Mann, 2009 www.littlebrownchurch.com
The New York Times, "American Treasures" 9/2001 by Sarah Ferrell
Chicago Tribune "Tiny Iowa Church Hosts 57th Marriage" by Amy Lorentzen 7/2009 Village of Bradford Historical Museum and Historical Society
Dr. William S. Pitts, Chicasaw County, Iowa History

Faith of A Quaker Miller:
Perry Hayden's Journal, September 30, 1940
The Lasting Impact of Perry Hayden's Life by John Hayden, 10/8/06Michigan
Hayden Hi-Lites published by Hayden Flour Mills, 7/26/1940, Benson Ford Research Center, Henry Ford Museum, Dearborn, MI Accession 285, Box 2279, "Ford Farms Macon 1939"
Author's extensive archives from Benson Ford Research Center and other sources
Martha Hayden Woodward taped interview 2/2004
Perry Hayden's Journal, 6/12/45, ""Henry Ford
Perry Hayden's Journal, 3/29/45, "Estimated cost of the 6th Year"
God is my Landlord by Raymond Jeffreys
"Faith of a Quaker Miller" is excerpted from Brenda Ervin's book about Dynamic Kernels, in 2011

The Old Rugged Cross Historic Hymn Site
Texas Baptists Top Tunes, Bread On The Water Magazine by June Cunningham source: International
"George Beverly Shea: Then Sings My Soul" DVD can be obtained through: www.gaither.com
Christian Ministries Magazine "The Answer",
"Giving Barns New Life," by author/historian Jerry Apps, pg. 4
Michigan Historical Site Marker: The Old Rugged Cross Historical Site
Christian History International
Email interview with Marta Dodd 5/29/09
Des Moines Registrar by "Famous Iowans" by writer Tom Longden
Michigan Historic Site Marker, "Methodist Episcopal Church"
Warsaw Times Union, 12/19/55, "Homer A Rodeheaver Dies"
Ideal Magazine "The Old Rugged Cross" by Robert Cushman Hayes
"Michigan Couple Restoring Birthplace of 'Old Rugged Cross', Edward Hoogterp, Baptist Standard 4/2001
The Old Rugged Cross Foundation, Inc web site: www.theoldruggedcross.org

Ryman & Rev. Sam P. Jones
The Ryman History, Nashville Post
Rev. Samuel Porter Jones, 1847-1906 by Chantel Parker, originally published in "150 Years of Cartersville,
Ryman Auditorium History, www.ryman.com
Daniel E. Sutherland, The Expansion of Everyday Life, University of Arkansas Press, pg. 94
Sermons by Samuel P. Jones, Cartersville Georgia, preached in St. Louis, 1886

God's Work & The American Barn
The Autobiography of Charles Finney by Helen Wessel, Bethany House, pg. 69
The First Presbyterian Church of Watertown, New York,
Parker, Tenney and White: pg. 266, silos
The Archaeological Encyclopedia of the Holy Land by Avraham Negev (not misspelled) pg. 142
The Bible Almanac, pg. 194
U. S. News & World Reports
The Life of James Abram Garfield by William r. Balch, 1881, pg. 63
Author Interview with Pastor Jean Richards Tulip, 2003, 2005, to see the Good Shepherd Mural log on to our web site at www.ruralamericabooks.org
North Carolina Barn Quilt Reference credited to Stephanie V. Slavin
Billy Frank Graham: Just As I Am", Sharing Our Heritage, pg. 17

Our Sacred Sanctuaries
King Memorial Baptist Church Documents
"North Dakotans seek salvation for churches," by Judy Keen, USA Today 11/2007
Preservation North Dakota, Preserving Prairie Places web site
American Profile Magazine, "Saving Sacred Places" by Marti Attoun
North Dakotans seek salvation for churches," by Judy Keen, USA Today 11/2007
Billy Graham God's Ambassador, pg. 35, The Florida Times-Union by Jeff Brumley "Billy Graham Turns 90: Palatka reveres an enduring ministry" 9/7/08
Peniel Baptist Church spokesperson and Peniel Baptist Church Spokesperson
Historic St. Mary's Mission web site, courtesy Colleen Meyer, Director
Montana History
San Miguel Mission Library of Congress "Historic American Building Survey," 1934
San Miguel Mission, Wikipedia, Santa Fe National Park Records, Santa Fe Visitors The Beauty of America, Tumacacori National Historical Park Commemorates Arizona's Oldest Spanish Mission by Bob Janiskee; National Historical Park Service
Mission Library, Juan de Torquemada 1834-1849, Englehardt (1922)

Faith on the Farm:
The Story of Oklahoma by Lon Tinkle, Landmark Books,
Will Rodgers: A Biography by Ben Yagoda, pg. 19
Oral Roberts: An American Life by David Edwin Harrell, Jr., Indiana University Press, pg. 9 & 12
Expect A Miracle; My Life and Ministry, Oral Roberts an Autobiography, pg. 4
Oral Roberts: An American Life by David Edwin Harrell, Jr., Indiana University Press, pg. 15
Expect A Miracle; My Life and Ministry, Oral Roberts An Autobiography, pg. 11
Oral Roberts: An American Life by David Edwin Harrell, Jr., Indiana University Press, pg. 15
Expect A Miracle; My Life and Ministry, Oral Roberts An Autobiography,
Still Doing the Impossible by Oral Roberts, pg. 187-189

TBN Interview, Behind the Scenes by Dr. Paul Crouch, Sr. & Paul Crouch Jr, August 2009
My Heart was in My Hands, Miracles, Oral Roberts Ministry, Vol.2, no. 3, 2009
When You See The Invisible You Can Do The Impossible
Dr. Paul Crouch, Behind the Scenes with Paul Crouch, Jr., TBN Interview with Dr. Oral Roberts

Village Missions of North America:
The Stock of Rural US America Churches by Gary Farley, File 1069, pg. 1
Nebraska Star Journal report, 2005
Ranching Realities by Joanna Nasar, American Cowboy Magazine, June/July 2008, pg. 40
Keeping the Wild in the West by Tom Arrandale, Governing Magazine, 10/2006, pg. 60
Successful Farmer's Magazine 2002 Reader's Survey
Rural Churches Grapple with a Pastor Exodus, David Van Bema/Crookston, 1/29/09 Time Magazine
Village Missions: A View from the Country by Brian Wechsler, pg. 4, Fall 2008
Pastor Larry Shetenhelm, Jennings Community Church, Interview 7/09
Village Mission pastor's interviews summation July/August 2009
From Six-Mile to Thirty-Mile Church Fields by Gary Farley, Vision, 1992
Village Missions: www.theyarenotforgotten.org
Author Interview with Pastor Larry Shetenhelm 7/2009

Farm Rescue:
Do Unto Others: Help Thy Neighbor by Micheal Haederle, Purpose Driven Connection, 2009
Farm Rescue Web Site www.farmrescue.org
Author's archives: interviews with farmers and descendants of farm families

America's Cowboy Churches:
Author Archives: Email Interview with Linda Beaumont, May 2009
Chisholm Trail Heritage Center: The Beginning of the Chisholm Trail
Author interview July 2009 with Jim Ballard web site: www.jimballard/namb.org
Cowboy Roams Idaho, planting churches by Mickey Noah, 6/2009, Southern Baptist Press
Epoch" Church on the Range publication by the NAMB, Southern Baptist Convention
Assembly of God: "Cowboy Church Set to Become Must See TV for Millions of Viewers"
Author's Interview with Susie Luchsinger, March 17 & 21, 2009
Luchsinger, Weaver bring Cowboy Church to Lazy E March 8th" by Ted Harbin, 2/26/09
Cowboy Church Set to Become Must See TV for Millions of Viewers," Assembly of God News
American Cowboy Magazine "The West: America's New Melting Pot" by Joanna Nasar, Tom Wilmes
Orlando-area Cowboy Church is at home on the range," by Jeff Kunerth, Orlando Sentinel, 2009
Church on the Range: Light in a Barn" by North American Mission Board

Barn is Cowboy Church in Lee County by Galen Holley, Northeast Mississippi Daily, 4/2009
Come as you are to the Cowboy Church-manure on books OK," by Barbara Hijek, SunSentinel writer
Author Interview with Jeff Smith, 2009 www.cowboycc.org
For more web site information log on to our web site at: www.ruralamericabooks.com

The Dineh, the Navajo's
The Indian Frontier of the American West, 1846-1890 by Robert M. Utley, pg. 84
Interview with Lola Yassie by phone, 2006, 2008
Under Sacred Ground: A History of Navajo Oil by Kathleen P. Chamberlain
Authors Interview with Martin & Christine Kirk and Lola Yassie conducted in 2007 & 2008
Interview conducted by Martin Kirk for the author of Lola Yassie, 2008
U S Senate Committee on Indian Affairs and the Senate Judiciary Committee September 17, 1997
Navajo Times, December 2007
Kirk's missionary letters July-October 2007
Author Interviews with Pastor Mike Calvin, Flagstaff Mission to the Navajo, 2009
Author Interviews with Pastor Scott Franklin, Flagstaff Mission to the Navajo, 2009

Bibliography

Cartwright, Peter, Autobiography of Peter Cartwright, Abingdon Press, 1984
Chamberlain, Kathleen, Under Sacred Ground: A History of Navajo Oil, ©2000
Cornwel, Patricia, The Ruth Bell Graham Story: A Time for Remembering, 1983, Davis, William C., The American Frontier, Smithmark Books, 1992
DeMoss Foundation, Rebirth of America, 1986
Engelhardt, Zephyrin, O. F. M. (1922) San Juan Capistrano Mission, Standard Publishing
Finney, Charles. G., The Autobiography of Charles G. Finney, The Life Story of America's Greatest Evangelist In His Own Words, Bethany House, 1977
Graham, Billy, Billy Graham: God's Ambassador
Graham, Billy, Just As I Am: The Autobiography of Billy Graham, HarperCollins, 2000
Hagin, Kenneth, Jr., From A Pastor's Heart, Kenneth Hagin Miinistries, 2000
Hayne, Coe, Baptist Trail-Makers of Michigan, Judson Press, 1936
Heilman, Grant, Farm, Abbeville Press, New York, 1988
Jakes, T. D., Mama Made the Difference ©2008 by T. D. Jakes, G. P. Putnam & Sons Jackson, Missouri Library, courtesy "History of Old McKendree Chapel
Jones, William J., Christ in the Camp, c. 1986, Sprinkle Publications
Lamb, Joni, Surrender All, Waterbrook Press, 2008
Larkin, Jack, The Reshaping of Everyday Life, 1790-1840, Harper, 1989
Loveland, Anne C., Southern Evangelicals and the Social Order, 1800-1860, c,1980
McEwen, Jerry, Room at the Inn: The William Tiny Zehnder Story
Murdock, Dr. Mike, The Wisdom Commentary of Mike Murdock, Copyright 2002 by MIKE MURDOCK, All publishing rights belong exclusively to Wisdom International. Published by The Wisdom Center, The Wisdom Center, 4051 Denton Highway, Fort Worth, Texas 76117 www.thewisdomcenter.tv
Murdock, Dr. J. E., Truth Out In The Open, Copyright 2003 Dr. J. E. Murdock, Rights belong exclusively to Wisdom International, The Wisdom Center, 4051 Denton Highway, Fort Worth, Texas 76117 (817-798-0300) www.thewisdomcenter.tv
Noll, Mark, The Rise of Evangelicalism, c.2003, Intervarsity Press
Sankey, Ira D., Dwight Lyman Moody's Life Work and Latest Sermons (As Delivered by the Great Evangelist) ©1901
Sharon Presbyterian Church "Sharing Our Heritage 1831-2006
Sherrill, John & Elizabeth, Demos Shakarian's "The Happiest People on Earth"
Sloane, Eric, America, c. 1982, Harper & Row
Stowe-Hambrick, Charles E., Charles G. Finney and the Spirit of American Evangelicalism, 1996

Sutherland, Daniel, E., The Expansion of Everyday Life 1860-1876
Roberts, Oral, Oral Roberts: Still Doing the Impossible, Destiny Image, c. 2002
Roberts, Oral, When You See the Invisible You Can Do The Impossible, c. 2002
Taylor, Alan, American Colonies: The Settling of North America, Penguin Books, 2001
Terrell, Bob, The Legacy of Buck & Dottie Rambo, StarSong Publishing, 1992
Tucker, Frank C., Old McKendree Chapel, Concord Publishing, 1984
Utley, Robert M., The Indian Frontier of the America West, 1984
Wikipedia: Documentation of San Juan Capistrano Mission
Wilkins, Teresa J., Patterns of Exchange, University of Oklahoma Press, 2008
White, Howard & Ruth, Mecklenburg: The Life and Times of a Proud People, c.
Zehnder, Herman F., Teach My People the Truth: The Story of Frankenmuth, Michigan

Other Books by Brenda Ervin

Barns of Michigan

Memories Of A Michigan Country Girl

A Home for Isabel: A Story of Faith, Hope & Courage

**Michigan Farms & Farm Families:
Farm Markets, Ag-Tourism & Barn Preservation**

*Perry Hayden & Henry Ford Farms:
The Great Wheat Experiment, 1940-1946*
(Coming in 2011)

Rural America Books

www.ruralamericabooks.org

Preserving & Publishing America's Rural Landscape

Rural America Books
PO Box 449
Hartland, MI 48353